FUZZY NEURAL NETWORKS FOR REAL TIME CONTROL APPLICATIONS

FUZZY NEURAL NETWORKS FOR REAL TIME CONTROL APPLICATIONS

Concepts, Modeling and Algorithms for Fast Learning

ERDAL KAYACAN
MOJTABA AHMADIEH KHANESAR

ELSEVIER

AMSTERDAM • BOSTON • HEIDELBERG • LONDON
NEW YORK • OXFORD • PARIS • SAN DIEGO
SAN FRANCISCO • SINGAPORE • SYDNEY • TOKYO
Butterworth-Heinemann is an imprint of Elsevier

Butterworth Heinemann is an imprint of Elsevier
The Boulevard, Langford Lane, Kidlington, Oxford OX5 1GB, UK
225 Wyman Street, Waltham, MA 02451, USA

British Library Cataloguing in Publication Data
A catalogue record for this book is available from the British Library

Library of Congress Cataloging-in-Publication Data
A catalog record for this book is available from the Library of Congress

For information on all Butterworth Heinemann publications
visit our website at http://store.elsevier.com/

ISBN: 978-0-12-802687-8

Publisher: Joe Hayton
Acquisition Editor: Sonnini Yura
Editorial Project Manager: Mariana Kühl Leme
Editorial Project Manager Intern: Ana Claudia A. Garcia
Production Manager: Kiruthika Govindaraju
Marketing Manager: Louise Springthorpe
Cover Designer: Mark Rogers

DEDICATION

To Omar Khayyám (1048–1131)
who played with mathematics
not only in science but also in his poems.

CONTENTS

FOREWORD

When Nilesh Karnik and I developed type-2 fuzzy logic systems (T2FLSs) in the late 1990s we required the following *fundamental design requirement* [1]: When all sources of (membership function) uncertainty disappear a T2FLS must reduce to a type-1 (T1)FLS. The biggest problem we faced was how to go from a type-2 fuzzy set (T2FS) to a number—the defuzzified output of the T2FLS. Our approach was to do this in two steps [2]: (1) type reduction (TR), in which a T2FS is projected into a T1FS, and (2) defuzzification of that T1FS. Unfortunately, type reduction cannot be performed using closed-form formulas; it is done using iterative algorithms, e.g., enhanced KM Algorithms [3]. Iterative TR may not be a good thing to do in a real-time T2FLS, especially for fuzzy logic control, because of its inherent time delays, and not having closed-form formulas means no mathematical analyses of the T2FLS can be performed, something that is abhorred by those who like to do analyses.

Beginning in 2001 some courageous researchers proposed T2FLSs that bypassed TR and went directly to a defuzzified output for the T2FLS. All of their T2FLSs satisfied the above fundamental design requirement. I call these researchers "courageous" because they had the courage to challenge what we had done, and this is what research should be about.

Hongwei Wu and Mendel [4] were the first ones to do this when they developed minimax uncertainty bounds (WM UBs) for the end points of the type-reduced set. These results became the starting architecture for some of the applications of Hani Hagras and his students [5]. Next came Nie and Tan [6] (NT) who took the union of the fired-rule output sets and computed the output of the resulting T2FLS as the center of gravity of the average of its lower and upper membership functions (MFs). Biglarbegian et al. [7] (BMM) proposed three simplified architectures each of which combined the left and right ends of the firing intervals in different ways. In one of these architectures the output of the T2FLS is a weighted combination of two very simple T1FLSs, one associated with the left-end values of the firing intervals and the other with the right-end values of the firing intervals (this is sometimes called an *m*-*n* or BMM architecture) and, in another of these architectures the T2FLS is assumed to be a weighted average of the average of the lower and upper firing intervals, where the weights are the respective consequents of TSK rules. The latter is analogous to the NT architecture when it is applied directly to the firing intervals and could be called the BMM-NT architecture.

All of these direct defuzzification approaches bypassed type reduction; however, the WM UB architecture was too complicated to be used in analyses of T2FLS, whereas the NT, BMM, and BMM-NT architectures are simple enough so that they can be used in analyses. Biglarbegian et al. already did this in [7–9] for the BMM architecture. The authors of the present book have done it for the BMM-NT architecture, and are to be congratulated for demonstrating many kinds of rigorous analyses that can be performed for it.

Optimizing T2MF parameters by using some training data is very important. When Qilian Liang and I were doing this for IT2FLSs that included type reduction, around 2000 [10], we focused on gradient-based optimization algorithms (e.g., steepest descent). Computing partial derivatives for such T2FLSs is fraught with difficulties [11] because the two switch points that are associated with the type-reduced set change from one time point to the next and the EKM Algorithms that are used to compute those switch points require a reordering of a set of numbers in an increasing order, after which the original ordering must be restored so that correct derivatives are computed. During the past decade (or longer) T2 researchers have focused on all kinds of alternatives to using derivative-based optimization algorithms, many of which are biologically inspired [12]. ACO, PSO, QPSO, and GA are some examples. Another benefit to using such algorithms is that (in theory) they will not get trapped in local extrema, whereas a gradient-based algorithm will.

When type reduction is not in the architecture of a T2FLS computing partial derivatives is quite simple because there no longer are switch points that change from one time to the next and there is no reordering of any set of numbers. The authors of this book have recognized this and have provided both some derivative-based and derivative-free optimization algorithms. They are able to perform some very serious convergence/stability analysis for all of them, for some parameters in their BMM-NT architecture. They are to be congratulated for demonstrating that serious analyses can indeed be performed for a T2FLS.

In summary, this book will be of great value to those who believe it is important to simplify T2FLSs and to use modern optimization algorithms to tune the MF parameters of such systems.

Jerry M. Mendel
University of Southern California, Los Angeles, CA
July 24, 2015

REFERENCES

[1] J.M. Mendel, Uncertain Rule-Based Fuzzy Logic Systems: Introduction and New Directions, Prentice-Hall, Upper Saddle River, NJ, 2001.

[2] N.N. Karnik, J.M. Mendel, Q. Liang, Type-2 fuzzy logic systems, IEEE Trans. Fuzzy Syst. 7 (1999) 643-658.

[3] D. Wu, J.M. Mendel, Enhanced Karnik-Mendel algorithms, IEEE Trans. Fuzzy Syst. 17 (2009) 923-934.

[4] H. Wu, J.M. Mendel, Uncertainty bounds and their use in the design of interval type-2 fuzzy logic systems, IEEE Trans. Fuzzy Syst. 10 (2002) 622-639.

[5] C. Lynch, H. Hagras, V. Callaghan, Using uncertainty bounds in the design of an embedded real-time type-2 neuro-fuzzy speed controller for marine diesel engines, in: Proc. FUZZ-IEEE 2006, Vancouver, Canada, 2006, pp. 7217-224.

[6] M. Nie, W.W. Tan, Towards an efficient type-reduction method for interval type-2 fuzzy logic systems?, in: Proc. IEEE FUZZ Conference, Paper # FS0339, Hong Kong, China, June 2008.

[7] M. Biglarbegian, W.W. Melek, J.M. Mendel, Stability analysis of type-2 fuzzy systems, in: Proc. of FUZZ-IEEE 2008, Hong Kong, China, June 2008, pp. 947-953.

[8] M. Biglarbegian, W.W. Melek, J.M. Mendel, On the stability of interval type-2 TSK fuzzy logic control systems, IEEE Trans. Syst. Man Cybernet. B Cybernet. 40 (2010) 798-818.

[9] M. Biglarbegian, W.W. Melek, J.M. Mendel, On the robustness of type-1 and interval type-2 fuzzy logic systems in modeling, Informat. Sci. 181 (2011) 1325-1347.

[10] Q. Liang, J.M. Mendel, Interval type-2 fuzzy logic systems: theory and design, IEEE Trans. Fuzzy Syst. 8 (2000) 535-550.

[11] J.M. Mendel, Computing derivatives in interval type-2 fuzzy logic systems, IEEE Trans. Fuzzy Syst. 12 (2004) 84-98.

[12] O. Castillo, P. Melin, Optimization of type-2 fuzzy systems based on bio-inspired methods: a concise review, Informat. Sci. 205 (2012) 1-19.

PREFACE

This book presents the basics of FNNs, in particular T2FNNs, for the identification and learning control of real-time systems. In addition to conventional parameter tuning methods, e.g., GD, SMC theory-based learning algorithms, which are simple and have closed forms, their stability analysis are also introduced. This book has been prepared in a way that can be easily understood by those who are both experienced and inexperienced in this field. Readers can benefit from the computer source codes for both identification and control purposes that are given at the end of the book.

There are number of books in the area of FLSs and FNNs. However, this book is more specific in several aspects. First of all, whereas so many books focus on the theory of type-1 and type-2 FLCs, we give more details on the parameter update algorithms of FNNs and their stability analysis. Second, the emphasis here is on the SMC theory-based learning algorithms for the training of FNNs, because we think these algorithms are the simplest and most efficient methods when compared to other algorithms, e.g., the GD algorithm. Last but not least, this book is prepared from the view of the identification and control of real-time systems, which makes it more practical.

The fuzzy logic principles were used to control a steam engine by Ebraham Mamdani of University of London in 1974. It was the first milestone for the fuzzy logic theory. The first industrial application was a cement kiln built in Denmark in 1975. In the 1980s, Fuji Electric applied fuzzy logic theory to the control a water purification process. As a challenging engineering project, in 1987, the Sendai Railway system that had automatic train operation FLCs since from 1987, not many books are available in the market as a reference for real-time systems. This book aims at filling this gap.

In Chapter 1, we summarize the basic mathematical preliminaries for a better understanding of the consecutive chapters. The given materials include the notations, definitions and related equations.

In Chapter 2, we introduce the concepts of type-1 fuzzy sets and T1FLCs. While Boolean logic results are restricted to 0 and 1, fuzzy logic results are between 0 and 1. In other words, fuzzy logic defines some intermediate values between sharp evaluations like absolute true and absolute false. That means fuzzy sets can handle concepts we commonly meet in daily life, like *very old*, *old*, *young*, and *very young*. Fuzzy logic is more like human thinking because it is based on degrees of truth and uses linguistic variables.

In Chapter 3, we introduce the basics of type-2 fuzzy sets, type-2 fuzzy MFs, and T2FLCs. There are two different approaches to FLSs design: T1FLSs and T2FLSs. The latter is proposed as an extension of the former with the intention of being able to model the uncertainties that invariably exist in the rule base of the system. In type-1 fuzzy sets, MFs are totally certain, whereas in type-2 fuzzy sets MFs are themselves fuzzy. The latter case results in the fact that the antecedents and consequents of the rules are uncertain.

In Chapter 4, type-1 and type-2 TSK fuzzy logic models are introduced. The two most common artificial intelligence techniques, fuzzy logic and ANNs, can be used in the same structure simultaneously, namely *FNNs*. The advantages of ANNs such as learning capability from input-output data, generalization capability and robustness and the advantages of fuzzy logic theory such as using expert knowledge are harmonized in FNNs. Instead of using fuzzy sets in the consequent part (like in Mamdani models), the TSK model uses a function of the input variables. The order of the function determines the order of the model, e.g., zeroth-order TSK model, first-order TSK model, etc.

In Chapter 5, we briefly discuss a multivariate optimization technique, namely the GD algorithm, to optimize a nonlinear unconstrained problem. In particular, the referred optimization problem is a cost function of a FNN, either type-1 or type-2. The main features, drawbacks and stability conditions of these algorithms are elaborated. Given an initial point, if an algorithm tries to follow the negative of the gradient of the function at the current point to be able to reach a local minimum, we face the most common iterative method to optimize a nonlinear function: the GD method.

In Chapter 6, the EKF algorithm is introduced to optimize the parameters of T2FNNs. The basic version of KF is an optimal linear estimator when the system is linear and is subject to white uncorrelated noise. However, it is possible to use Taylor expansion to extend its applications to nonlinear cases. Finally, the decoupled version of the EKF is also discussed, which is computationally more efficient than EKF to tune the parameters of T2FNNs.

In Chapter 7, in order to deal with nonlinearities, lack of modeling, several uncertainties and noise in both identification and control problems,

SMC theory-based learning algorithms are designed to tune both the premise and consequent parts of T2FNNs. Furthermore, the stability of the learning algorithms for control and identification purposes are proved by using appropriate Lyapunov functions. In addition to its well-known feature of being robust, the most significant advantage of the proposed learning algorithm for the identification case is that the algorithm has a closed form, and thus it is easier to implement in real-time when compared to the other existing methods.

In Chapter 8, a novel hybrid training method based on continuous version of PSO and SMC theory-based training method for T2FNNs is proposed. The approach uses PSO for the training of the antecedent part of T2FNNs, which appear nonlinearly in the output of T2FNNs, and SMC theory-based training method for the training of the parameters of their consequent part. The use of PSO makes it possible to lessen the probability of entrapment of the parameters in a local minima while proposing simple adaptation laws for the parameters of the antecedent part of T2FNNs when compared to the most popular approaches like GD, LM and EKF. The stability of the proposed hybrid training method is proved by using an appropriate Lyapunov function.

In Chapter 9, an attempt is made to show the effect of input noise in the rule base numerically in a general way. There exist number of papers in literature claiming that the performance of T2FLSs is better than their type-1 counterparts under noisy conditions. They attempt to justify this claim by simulation studies only for some specific systems. However, in this chapter, such an analysis is done independent of the system to be controlled. For such an analysis, a novel type-2 fuzzy MF (elliptic MF) is proposed. This type-2 MF has certain values on both ends of the support and the kernel and some uncertain values for the other values of the support. The findings of the general analysis in this chapter and the aforementioned studies published in literature are coherent.

In Chapter 10, the learning algorithms proposed in the previous chapters (GD-based, SMC theory-based, EKF and hybrid PSO-based learning algorithms) are used to identify and predict two nonlinear systems, namely Mackey-Glass and a second-order nonlinear time-varying plant. Several comparisons are done, and it has been shown that the proposed SMC theory-based algorithm has faster convergence speed than the existing methods such as the GD-based and swarm intelligence-based methods. Moreover, the proposed learning algorithm has an explicit form, and it is easier to implement than other existing methods. However, for offline algorithms for which computation time is not an issue, the hybrid training method based on PSO and SMC theory may be a preferable choice.

In Chapter 11, three real-world control problems, namely anesthesia, magnetic rigid spacecraft and tractor-implement system, are studied by using SMC theory-based learning algorithms for T2FNNs. For all the systems, the FEL scheme is preferred in which a conventional controller (PD, etc.) works in parallel with an intelligent structure (T1FNNs, T2FNN, etc.). The proposed learning algorithms have been shown to be able to control these real-world example problems with a satisfactory performance. Note that the proposed control algorithms do not need a priori knowledge of the system to be controlled.

Potential readers of this book are expected to be undergraduate and graduate students, engineers, mathematicians and computer scientists. Not only can this book be used as a reference source for a scientist who is interested in FNNs and their real-time implementation but also as a course book on FNNs or artificial intelligence in master or doctorate university studies. We hope this book will serve its main purpose successfully.

Erdal Kayacan
(Nanyang Technological University, Singapore)

Mojtaba Ahmadieh Khanesar
(Semnan University, Iran)
June 2015

ACKNOWLEDGMENTS

We would like to acknowledge our families for their support and patience, without whom this book would have been incomplete.

LIST OF ACRONYMS/ABBREVIATIONS

ANN	artificial neural network
BP	back propagation
BIBO	bounded input bounded output
BIS	bispectral index
CPSO	continous-time particle swarm optimization
DEKF	decoupled extended Kalman filter
DCN	distortion caused by noise
EKF	extended Kalman filter
EPSC	extended prediction self-adaptive controller
FEL	feedback error learning
FOU	footprint of uncertainty
FLC	fuzzy logic controller
FLS	fuzzy logic system
FNN	fuzzy neural network
FPGA	field-programmable gate array
GD	gradient descent
HR	heart rate
HCPSO	higher-order continous-time particle swarm optimization
KF	Kalman filter
LM	Levenberg-Marquardt
LBM	lean body mass
MPC	model predictive control
MSE	mean-squared error
MF	membership function
MIMO	multiple input multiple output
MISO	multiple input single output
ODE	ordinary differential equation
PD	proportional-derivative
PID	proportional-derivative-integral
PSO	particle swarm optimization
RMSE	root mean-squared error
SISO	single input single output

SNR	signal-to-noise ratio
SMC	sliding mode control
SSE	squared of the error
TSK	Takagi-Sugeno-Kang
T1FLC	type-1 fuzzy logic controller
T1FLS	type-1 fuzzy logic system
T1FNN	type-1 fuzzy neural network
T2FLC	type-2 fuzzy logic controller
T2FLS	type-2 fuzzy logic system
T2FNN	type-2 fuzzy neural network
VLSI	very-large-scale integration

CHAPTER 1

Mathematical Preliminaries

Contents

Abstract

This chapter summarizes the basic mathematical preliminaries for a better understanding of the consecutive chapters. The given materials include the notations, definitions and related equations.

Keywords

Matrix, Matrix inversion, Functions, Taylor expansion, Gradient, Hessian matrix, Stability analysis, Lyapunov function

1.1 INTRODUCTION

Design, optimization and parameter tuning of FNNs require fundamental knowledge about matrix theory, linear algebra, function approximation, partial derivatives, nonlinear programming, state estimation and nonlinear stability analysis. Although this chapter summarizes a few fundamental mathematical preliminaries that will allow the reader to follow the consecutive chapters easier, they are selective, and serve only as a reference for the notations and theories used in this book. For a more detailed explanation of the theories, the reader is encouraged to refer to other textbooks.

1.2 LINEAR MATRIX ALGEBRA

A matrix is a rectangular array of elements that are usually numbers or functions arranged in rows and columns. Let A be a $m \times n$ matrix given as follows [1]:

Fuzzy Neural Networks for Real Time Control Applications
http://dx.doi.org/10.1016/B978-0-12-802687-8.00001-3

$$A = \begin{bmatrix} a_{11} & a_{12} & \cdots & a_{1n} \\ a_{21} & a_{22} & \cdots & a_{2n} \\ \vdots & \vdots & & \vdots \\ a_{m1} & a_{m2} & \cdots & a_{mn} \end{bmatrix} \tag{1.1}$$

which has m rows and n columns, and a_{ij} is the element of the matrix A in ith row and jth column. If $m = n$, the matrix A is called a square matrix. If a_i is a $m \times 1$ matrix, it is called a column vector. Similarly, if it is a $1 \times n$ matrix, it is called a row vector. If the sizes of the matrices A and B are $n \times m$ and $m \times p$, respectively, then their matrix multiplication is defined as follows:

$$AB = \left[\sum_{k=1}^{m} a_{ik} b_{kj} \right] \tag{1.2}$$

To be able to multiply two matrices, the number of the columns of the first matrix must be equal to the number of the rows of the second matrix.

The transpose of the matrix A is represented by A^{T} and is obtained by rewriting all rows of the matrix as its columns and all its columns as its rows. Among number of different properties of the transpose operator, the most two frequently used ones are as follows:

$$(A^{\mathrm{T}})^{\mathrm{T}} = A \tag{1.3}$$

$$(AB)^{\mathrm{T}} = B^{\mathrm{T}} A^{\mathrm{T}} \tag{1.4}$$

An $n \times n$ matrix A is called symmetric if $A^{\mathrm{T}} = A$, and it is called skew-symmetric if $A^{\mathrm{T}} = -A$. Any matrix can be rewritten as a summation of its symmetric and skew-symmetric matrices as follows:

$$A = \frac{1}{2}(A + A^{\mathrm{T}}) + \frac{1}{2}(A - A^{\mathrm{T}}) \tag{1.5}$$

where $\frac{1}{2}(A + A^{\mathrm{T}})$ is the symmetric and $\frac{1}{2}(A - A^{\mathrm{T}})$ is the skew-symmetric part.

If the elements $a_{ij} = 0, \ \forall \, i \neq j$ for a $n \times n$ matrix A, then the matrix A is called a diagonal matrix and can be represented by:

$$A = \mathrm{diag}(a_{11}, a_{22}, \ldots, a_{nn}) \tag{1.6}$$

Furthermore, if $a_{ii} = 1,\ i = 1, \ldots, n$, the matrix is called an identity matrix and is represented by I_n. The identity matrix has the following property:

$$AA^{-1} = I, \quad \forall A \in \mathbb{R}^n \tag{1.7}$$

where A^{-1} is called the inverse of the matrix A. However, if the matrix A is not a square matrix, then a unique matrix $A^{\dagger}$ is called the pseudo-inverse of the matrix A provided that it satisfies the following conditions:

1. $AA^{\dagger}A = A$
2. $A^{\dagger}AA^{\dagger} = A^{\dagger}$
3. $(AA^{\dagger})^{\mathrm{T}} = AA^{\dagger}$
4. $(A^{\dagger}A)^{\mathrm{T}} = A^{\dagger}A$

If the matrix A is square and non-singular, then the pseudo-inverse of A is equal to its inverse, i.e., $A^{\dagger} = A^{-1}$.

There exist different lemmas for the inversion of a matrix, one of which is as follows:

Lemma 1.1 (Matrix Inversion Lemma [2]). *Let A, C, and $C^{-1} + DA^{-1}B$ be nonsingular square matrices. Then $A + BCD$ is invertible, and*

$$(A + BCD)^{-1} = A^{-1} - A^{-1}B(C^{-1} + DA^{-1}B)^{-1}DA^{-1}$$

Proof. The following can be obtained by using direct multiplication:

$$\begin{aligned}(A + BCD) \times \left(A^{-1} - A^{-1}B(C^{-1} + DA^{-1}B)^{-1}DA^{-1}\right) \\ &= I + BCDA^{-1} - B(C^{-1} + DA^{-1}B)^{-1}DA^{-1} \\ &\quad - BCDA^{-1}B(C^{-1} + DA^{-1}B)^{-1}DA^{-1} \\ &= I + BCDA^{-1} - BC(C^{-1} + DA^{-1}B)(C^{-1} + DA^{-1}B)^{-1}DA^{-1} \\ &= I\end{aligned}$$

□

Let $A \in \mathbb{R}^{n \times n}$ be a square matrix. The scalar λ_i is called the eigenvalue of the matrix A if it satisfies the following equation [3]:

$$Av_i = \lambda_i v_i \tag{1.8}$$

and v_i is called its corresponding eigenvectors.

The condition number of a matrix A is defined as:

$$\text{cond}(A) = \left| \frac{\lambda_{\text{max}}(A)}{\lambda_{\text{min}}(A)} \right| \tag{1.9}$$

where $\lambda_{\text{max}}(A)$ and $\lambda_{\text{min}}(A)$ represent the largest and the smallest eigenvalues of the matrix A, respectively. Apparently, the condition number is greater than 1. A large value for the condition number indicates that the matrix A is ill-conditioned and may suffer from numerical instability specially when it comes to computing its inverse.

1.3 FUNCTION

A real-valued function $F(x)$ may be defined in a vector form as follows:

$$F(x) = [F_1, F_2, \ldots, F_m], \ F : \mathbb{R}^n \to \mathbb{R}^m \tag{1.10}$$

where F_i, $i = 1, \ldots, m$ the elements of the matrix F are real-valued functions of real numbers. If $F(x)$ is a scalar matrix, its gradient is defined as follows:

$$\nabla_x F(x) = \frac{\partial F}{\partial x} = \left[\frac{\partial F}{\partial x_1}, \frac{\partial F}{\partial x_2}, \ldots, \frac{\partial F}{\partial x_n} \right]^{\text{T}} \tag{1.11}$$

The gradient of the function $F(x)$ is a column vector.

The Hessian matrix of the function $F(x)$ is defined as follows [3]:

$$H(x) = \nabla^2 F(x) = \frac{\partial}{\partial x} \left[\frac{\partial F(x)}{\partial x} \right]^{\text{T}} \tag{1.12}$$

and is calculated as:

$$H(x) = \begin{bmatrix} \frac{\partial^2 F}{\partial x_1^2} & \frac{\partial^2 F}{\partial x_1 \partial x_2} & \cdots & \frac{\partial^2 F}{\partial x_1 \partial x_n} \\ \frac{\partial^2 F}{\partial x_2 \partial x_1} & \frac{\partial^2 F}{\partial x_2^2} & \cdots & \frac{\partial^2 F}{\partial x_2 \partial x_n} \\ \vdots & \vdots & \ddots & \vdots \\ \frac{\partial^2 F}{\partial x_n \partial x_1} & \frac{\partial^2 F}{\partial x_n \partial x_2} & \cdots & \frac{\partial^2 F}{\partial x_n^2} \end{bmatrix} \tag{1.13}$$

If $F(x)$ is a vector function as:

$$F(x) = [F_1(x),\ F_2(x), \ldots, F_m(x)]\,,\ \ \mathbb{R}^n \rightarrow \mathbb{R}^m \tag{1.14}$$

then its Jacobian matrix is defined as follows:

$$J(x) = \frac{\partial F(x)}{\partial x} = \begin{bmatrix} \frac{\partial F_1}{\partial x_1} & \frac{\partial F_1}{\partial x_2} & \cdots & \frac{\partial F_1}{\partial x_n} \\ \frac{\partial F_2}{\partial x_1} & \frac{\partial F_2}{\partial x_2} & \cdots & \frac{\partial F_2}{\partial x_n} \\ \vdots & \vdots & \ddots & \vdots \\ \frac{\partial F_m}{\partial x_1} & \frac{\partial F_m}{\partial x_2} & \cdots & \frac{\partial F_m}{\partial x_n} \end{bmatrix} \tag{1.15}$$

Let $g(.)\,:\,\mathbb{R}^r \rightarrow \mathbb{R}^m$ and $h(.)\,:\,\mathbb{R}^n \rightarrow \mathbb{R}^r$ be two functions and $F(.)$ is defined as follows:

$$F(x) = g(h(x)) \tag{1.16}$$

then we have:

$$\nabla F(x) = \nabla_h g(h(x)) \nabla_x h(x) \tag{1.17}$$

Let $F(x)$ be a real-valued and differentiable scalar function of the input vector $x \in \mathbb{R}^n$. The Taylor expansion of $F(x)$ about the point x_0 is:

$$F(x) = F(x_0) + \sum_{i=1}^{n} \frac{\partial F(x)}{\partial x_i} \Delta x_i + \frac{1}{2} \sum_{i=1}^{n} \sum_{j=1}^{n} \frac{\partial^2 F(x)}{\partial x_i \partial x_j} \Delta x_i \Delta x_j + \text{H.O.T.} \tag{1.18}$$

where $\Delta x_i = x_i - x_{0i}$ is the deviation of the variable x from x_0 in its ith dimension and H.O.T. represents the higher-order terms.

1.4 STABILITY ANALYSIS

In order to prove the stability of a nonlinear ordinary differential equation, the Lyapunov stability theory is the most powerful and applicable method. In order to investigate the stability of a system using this method, a positive energy-like function of the states on the trajectory of the differential equation is taken into account, which is called the Lyapunov function. A Lyapunov function decreases along the trajectory of the ODE.

Let the ordinary differential equation of the system be of the following form:

$$\dot{x} = F(x) \tag{1.19}$$

where $F(x) : \mathbb{R}^n \rightarrow \mathbb{R}^n$ and $x \in \mathbb{R}^n$ be the state vector of the system. The following theorem holds:

Theorem 1.1 (Stability of Continuous Time Systems [4]). *Let $x = 0$ be an equilibrium point and $D \in \mathbb{R}^n$ be a domain containing $x = 0$. Let $V : D \rightarrow \mathbb{R}$ be a continually differentiable function such that:*

$$V(0) = 0, \ \ and \ V(x) > 0 \ in \ D - \{0\} \tag{1.20}$$

and

$$\dot{V} \leq 0 \ in \ D \tag{1.21}$$

then $x = 0$ is stable. Moreover, if:

$$\dot{V} < 0 \ in \ D - \{0\} \tag{1.22}$$

then $x = 0$ is asymptotically stable.

It is also possible to investigate the stability of a discrete time difference equation using Lyapunov theory. Let the discrete time difference equation of a system be of the following form:

$$x_{k+1} = F(x_k) \tag{1.23}$$

where $F(x) : \mathbb{R}^n \rightarrow \mathbb{R}^n$ and $x \in \mathbb{R}^n$ be the state vector of the system. The following theorem holds:

Theorem 1.2 (Stability of Discrete Time Systems [5]). *Let $x = 0$ be an equilibrium point and $D \in \mathbb{R}^n$ be a domain containing $x = 0$. Let $V : D \rightarrow \mathbb{R}$ be a continually differentiable function such that*

$$V(0) = 0, \ \ and \ V(x) > 0 \ in \ D - \{0\} \tag{1.24}$$

then $x = 0$ is stable. Moreover, if:

$$\Delta V = V(x_{k+1}) - V(x_k) < 0 \ in \ D - \{0\} \tag{1.25}$$

then $x = 0$ is asymptotically stable.

1.5 SLIDING MODE CONTROL THEORY

SMC is a robust control design technique that can successfully be applied to linear and nonlinear dynamic systems in which the desired behavior of the system is defined as a sliding manifold. This control method has two modes: reaching (hitting) mode and sliding mode [4]. Reaching mode occurs when the states of the system are not on the sliding manifold. In this mode, the controller must be designed such that it drives the states of the system to the sliding manifold. When the states of the system reach the sliding manifold, the controller must maintain the states of the system on it. This mode is called sliding mode.

For a better understanding of SMC, the following example is given:

Consider a second-order nonlinear dynamic system as follows:

$$\begin{aligned} \dot{x}_1 &= x_2 \\ \dot{x}_2 &= f(x) + u \end{aligned} \tag{1.26}$$

where $f(x) : \mathbb{R}^2 \rightarrow \mathbb{R}$ is a nonlinear function of the states of the system. It is assumed that $f(x)$ is partially known and has the following form:

$$f(x) = f_n(x) + \Delta f(x) \tag{1.27}$$

where $f_n(x)$ is the nominal part of $f(x)$ that is considered to be known, and $\Delta f(x)$ is the unknown part of $f(x)$, which satisfies $|\Delta f(x)| < F$. The duty of the sliding mode controller is to design a control law to constrain the motion of the system to the manifold (in this case, a sliding line as in Fig. 1.1) $s = \lambda x_1 + x_2 = 0$. On this manifold, the motion of the system is governed by $\dot{x}_1 = -\lambda x_1$, which is the desired behavior of the system. The choice of $\lambda > 0$ guarantees that $x(t)$ tends to zero as t tends to infinity. The rate of convergence is controlled by the choice of λ. When the states of the system are on the manifold, the controlled trajectories of the system are independent of the dynamic of the system $f(x)$. For the time derivative of s, the following equation is obtained:

$$\dot{s} = \lambda \dot{x}_1 + \dot{x}_2 = \lambda x_2 + f(x) + u \tag{1.28}$$

A Lyapunov function can be defined as follows:

$$V = \frac{1}{2} s^2 \tag{1.29}$$

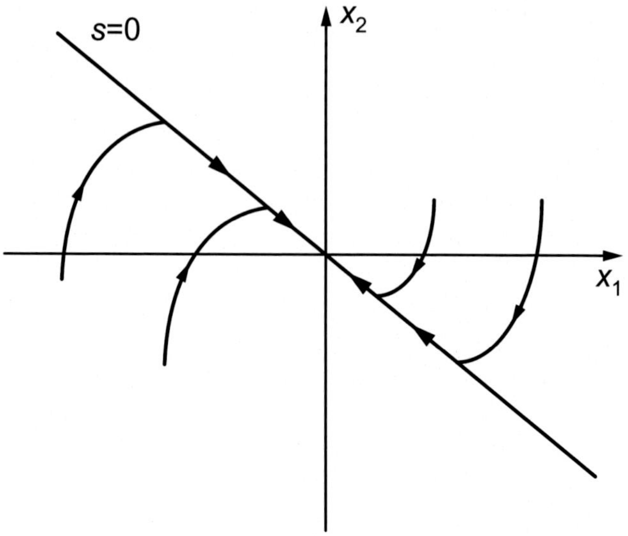

Figure 1.1 Typical phase portrait under SMC.

The time derivative of this function is as follows:

$$\dot{V} = s\dot{s} \tag{1.30}$$

Considering (1.28), the following equation is obtained:

$$\dot{V} = s(\lambda x_2 + f(x) + u) \tag{1.31}$$

and further:

$$\dot{V} = s(\lambda x_2 + f_n(x) + \Delta f(x) + u) \tag{1.32}$$

The control signal u is assumed to have two parts as follows:

$$u = u_n + u_r \tag{1.33}$$

where u_n represents the part of the control signal, which cancels the nominal part of the nonlinear function and u_r is a term to guarantee the robustness of the controller in the presence of $\Delta f(x)$. The nominal part of the control signal, u_n, is taken as follows:

$$\dot{V} = s(\Delta f(x) + u_r) \tag{1.34}$$

In order to have a finite time convergence of the manifold to zero, it is assumed that:

$$\dot{V} \leq -\eta|s| \tag{1.35}$$

and hence:

$$s(\Delta f(x) + u_r) < -\eta|s| \tag{1.36}$$

u_r is taken as follows:

$$u_r = -K\text{sgn}(s) \tag{1.37}$$

where sgn(s) is defined as follows:

$$\text{sgn}(s) = \begin{cases} 1, & s > 0 \\ -1 & s < 0 \\ 0 & s = 0 \end{cases} \tag{1.38}$$

Therefore:

$$|s||\Delta f(x)| - K|s| < -\eta|s| \tag{1.39}$$

and further:

$$-K|s| < -\eta|s| - F|s| \tag{1.40}$$

or equivalently:

$$\eta + F < K \tag{1.41}$$

Hence, the SMC, which guarantees the stability of the system, is as follows:

$$u = -\lambda x_2 - f_n(x) - K\text{sgn}(s) \tag{1.42}$$

A typical phase portrait of a second-order nonlinear dynamic system and its sliding manifold is depicted in Fig. 1.1.

As already mentioned earlier, SMC is a robust control algorithm that can control nonlinear dynamic systems in the presence of uncertainties. However, this controller suffers from several drawbacks:

1. The presence of sgn function makes the controller sensitive to noise. For instance, suppose that the value of s is equal to 10^{-6} and a small value of noise with amplitude of 10^{-5} with a different sign is added to s. This small value can vary the sign of u_r. Since K has a large value, it may greatly affect the control signal and mislead the system.
2. In theory, when the states of the system hit the manifold, the ideal SMC maintains them on it. However, in practice, there are some uncertainties in the system and there also exists a delay between the time that the sign of s changes and the controller switches. These reasons result in the states crossing the manifold. This event is repetitive and produces a zig-zag motion in the phase plane with a high frequency and is called *chattering* (see Fig. 1.2)

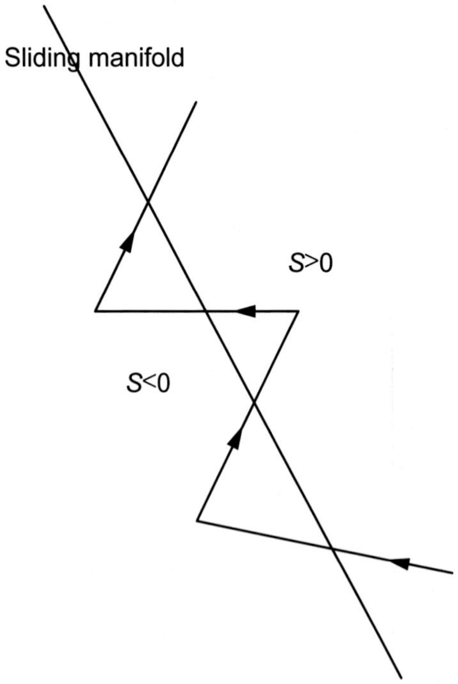

Figure 1.2 Chattering due to delay and uncertainty in control signal switching.

3. In the design of an SMC, it is required that the nominal value of the nonlinear functions of the system (e.g., f_n in our example) is known. Furthermore, the upper bound of the uncertainties in the system (e.g., $\Delta f(x)$ in our example) is also known.

In order to cope with these challenges, different methods have been proposed: classical approaches and intelligent methods. The classical approaches replace the sign function that usually exists in SMC with a smooth function. Furthermore, as was mentioned earlier, a priori knowledge about the nominal functions of the system is required in order to design an SMC. The adaptive approaches may be used to approximate these functions and hence lessen the need of SMC to priori knowledge about the system. However, the intelligent methods use neural networks, fuzzy logic and FNNs to cope with the challenging problems of SMC. Because of the proven general function approximation property, flexibility and capability of using human knowledge of FNNs, this structure is one of the most important structures to overcome the drawbacks of SMC [6–8]. The intelligent methods are known to cause less chattering in control signal and eliminate the knowledge needed of the dynamic model of the system.

1.6 CONCLUSION

The aim of this chapter is to make the next chapters easier to follow and understand. In order to accomplish this, a summary of basic mathematical relationships and equations that will be used in consecutive chapters is given. The reader is encouraged to refer to other textbooks for more detailed explanations and discussions.

REFERENCES

[1] A. Cochocki, R. Unbehauen, Neural Networks for Optimization and Signal Processing, John Wiley & Sons, Inc., New York, 1993.
[2] K.J. Åström, B. Wittenmark, Adaptive Control, Courier Corporation, Mineola, NY, 2013.
[3] X.-S. Yang, Engineering Optimization: An Introduction with Metaheuristic Applications, John Wiley & Sons, , 2010.
[4] H.K. Khalil, J. Grizzle, Nonlinear Systems, vol. 3, Prentice Hall, Upper Saddle River, NJ, 1996.
[5] M. Gupta, L. Jin, N. Homma, Static and Dynamic Neural Networks: From Fundamentals to Advanced Theory, John Wiley & Sons, New York, 2004.
[6] H.-Y.C. Lon-Chen Hung, Decoupled sliding-mode with fuzzy-neural network controller for nonlinear systems, Int. J. Approx. Reason. 46 (2007) 74–97.

[7] E. Kayacan, O. Cigdem, O. Kaynak, Sliding mode control approach for online learning as applied to Type-2 fuzzy neural networks and its experimental evaluation, IEEE Trans. Indust. Elect. 59 (9) (2012) 3510–3520.
[8] M. Khanesar, E. Kayacan, M. Reyhanoglu, O. Kaynak, Feedback error learning control of magnetic satellites using Type-2 fuzzy neural networks with elliptic membership functions, IEEE Trans. Cybernet. 45 (4) (2015) 858–868.

CHAPTER 2

Fundamentals of Type-1 Fuzzy Logic Theory

Contents

Abstract

While Boolean logic results are restricted to 0 and 1, fuzzy logic results are between 0 and 1. In other words, fuzzy logic, as a super set of conventional Boolean logic, defines some intermediate values between sharp evaluations like absolute true and absolute false. That means fuzzy sets can handle concepts we commonly face in daily life, like *very old*, *old*, *young* and *very young*. In this chapter, we introduce the concepts of type-1 fuzzy sets and T1FLCs.

Keywords

Fuzzy logic theory, Fuzzy logic control, Crisp sets, Fuzzy sets, Boolean logic

2.1 INTRODUCTION

Fuzzy theory was first proposed in 1965 by Professor Lotfi A. Zadeh at the University of California at Berkeley who introduced a set theory that operates over the range [0, 1]. The core of his theory was outlined in his seminal work entitled "Fuzzy Sets" in the journal *Information and Control* [1].

Fuzzy Neural Networks for Real Time Control Applications
http://dx.doi.org/10.1016/B978-0-12-802687-8.00002-5

While Boolean logic results are restricted to 0 and 1, fuzzy logic results are between 0 and 1. In other words, fuzzy logic defines some intermediate values between sharp evaluations like absolute true and absolute false. This means fuzzy sets can handle concepts we commonly face in daily life, like *very old*, *old*, *young* and *very young*. Fuzzy logic is more like human thinking because it is based on degrees of truth and uses linguistic variables.

Fuzzy logic deals with fuzzy sets whose elements have degrees of memberships. In other words, an element can be a member of more than one set associated with different membership values. For instance, in most Western countries, weekdays are from Monday to Friday; the weekend includes Saturday and Sunday. As an alternative to this Boolean logic, one can think that people start feeling the positive effect of a weekend on Friday. Thus, we may think that while Friday belongs to the set of "weekdays" with a membership value of 0.9, it belongs to the set of "weekend" with a membership value of 0.1. A similar logic can be constructed for Sunday too. This example illustrates that while the conventional logic deals with "truth of any statement," fuzzy logic concerns with "degree of truth."

Zadeh makes the following statement for the fuzzy logic:

> *"Fuzzy logic is a precise conceptual system of reasoning, deduction, and computation in which the objects of discourse and analysis are, or are allowed to be, associated with imperfect information. Imperfect information is information which in one or more respects is imprecise, uncertain, incomplete, unreliable, vague or partially true" [2].*

Fuzzy logic was probably not an acceptable concept for the scientists in the early 1960s, because it contained vagueness in the engineering field. In particular, for the Western countries, the concept of "fuzzy" had a negative connotation both in science and engineering. Science exists to get rid of "vagueness" in our daily life and make everything clear. So, why would we bring a concept such as "fuzzy" into science?

Even if there were many scientists who had considered fuzzy logic to be unscientific, since the 1970s, this approach to set theory has been widely applied to control systems. As a significant milestone, the principles of fuzzy logic were used to control a steam engine by Ebraham Mamdani of University of London in 1974 [3]. The first industrial application was a cement kiln built in Denmark in 1975. In the 1980s, Fuji Electric applied fuzzy logic theory to the control of a water purification process. As a challenging engineering project, in 1987, Sendai Railway system that had automatic train operation control was built with fuzzy logic principles in

Japan. Fuzzy control techniques were used in all the critical operations in the control of the train, such as accelerating, breaking, and stopping operations. In 1987, Takeshi Yamakawa used fuzzy control in an inverted pendulum experiment, which is a classical benchmark control problem. After these successful applications, not only the engineers but also the social scientists applied fuzzy logic to different areas. In today's technology, many companies use fuzzy logic in their engineering projects for air conditioners, video cameras, televisions, washing machines, bus time tables, decision-making systems, medical diagnoses, etc.

The classical model-based control theory, typically a PID controller, uses a mathematical model to define the relationship between the input and output. The most serious disadvantage of these controllers is that they usually assume the system to be linear or at least that it behaves as a linear system in some range. If an accurate mathematical model of a system is available, a conventional PID controller can make the performance of the system quite acceptable. However, in real-time industrial systems, it is often the case that there exist considerable difficulties in obtaining an accurate model. Even when the model is sufficiently accurate, there are many other uncertainties, e.g., due to the precision of the sensors, noise produced by the sensors, environmental conditions of the sensors and nonlinear characteristics of the actuators. Then, not only does the performance of the model-based approaches drastically decrease, but also the complexity of the controller design increases. In real-time applications, one solution is to tune the controller coefficients conservatively. Then, robustness comes at a price of a limited control performance beyond question. In such cases, a model-free approach is preferable. Fortunately, FLCs have the ability to control a system using some limited expert knowledge. In most cases, the design procedure of a FLC tries to imitate an expert or a skilled human operator. Besides, FLCs are low-cost implementations based on cheap sensors.

Although the concept of fuzzy logic and the concept of probability seem to be similar, they are quite different. While probability makes guesses about a certain reality, fuzzy logic does not make probability statements but represents membership in vaguely defined sets. For instance, if 0.5 is defined as a probability value for a person to be old, it can be said that there is a chance that he/she may be old. It is not known whether he/she is old or young. However, in fuzzy logic, if 0.5 is defined as the degree of membership in the set of young and old people, we have some knowledge about his/him and he/she is positioned in the middle of young and old people.

2.2 TYPE-1 FUZZY SETS

Fuzzy logic is considered much closer in spirit to human thinking and natural language. The way of human thinking is realized with MFs, which define how every point in the input space is mapped to a membership values space. The membership values in fuzzy sets are in the range of [0;1]. If an axiom is absolutely true, the membership value in fuzzy sets is 1. Similarly, if it is absolutely false, the membership value in fuzzy sets is 0. The output of a MF is called an antecedent (μ). While the input values for a MF are crisp, they are changed into fuzzy variables by these MFs.

Figures 2.1 and 2.2 show the difference between classical Boolean (or binary) logic and fuzzy logic. In Fig. 2.1, the MF is sharp-edged, which means a small change in an input value might cause a big change in the output values. According to this type of logic, any person shorter than 170 cm is considered to be short. However, in daily life, our way of thinking is completely different, but similar to the one in Fig. 2.2. In the latter approach, the MF is continuous and smooth and is more like human thinking. According to the green expert in Fig. 2.2, any person shorter than 150 cm is considered to be short, whereas any person taller than 190 cm is tall. Meanwhile, people who are taller than 150 cm but shorter than 190 cm belong to both the "short" and "tall" sets with specific membership values.

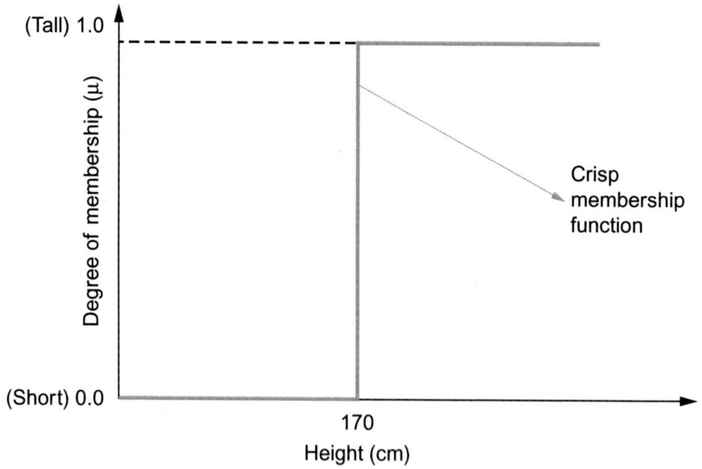

Figure 2.1 Possible description of the vague concept "tall" by a crisp set.

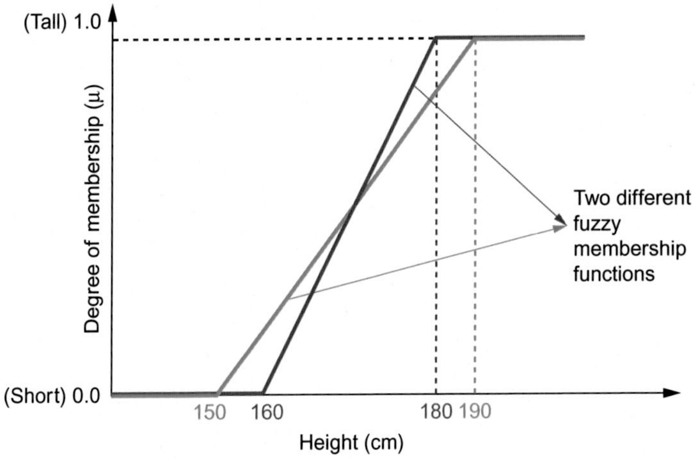

Figure 2.2 Possible description of the vague concept "tall" by a fuzzy set.

On the other hand, the red expert thinks in a slightly different way. His/her thresholds for being "short" and "tall" are 160 and 180 cm, respectively. The difference between these two experts brings the phenomena of "expert knowledge" to fuzzy logic theory. This feature is, of course, an additional degree of freedom.

A type-1 fuzzy set, A, which is in terms of a single variable, $x \in X$, may be represented as:

$$A = \{(x, \mu_A(x)) | \quad \forall x \in X\} \tag{2.1}$$

A can also be defined as:

$$A = \int_{x \in X} \mu_A(x)/x \tag{2.2}$$

where $\int$ denotes union over all admissible x.

As can be seen from Fig. 2.3, a type-1 Gaussian MF, $\mu_A(x)$, is constrained to be between 0 and 1 for all $x \in X$, and is a two-dimensional function. This type of MF does not contain any uncertainty. In other words, there exists a clear membership value for every input data point.

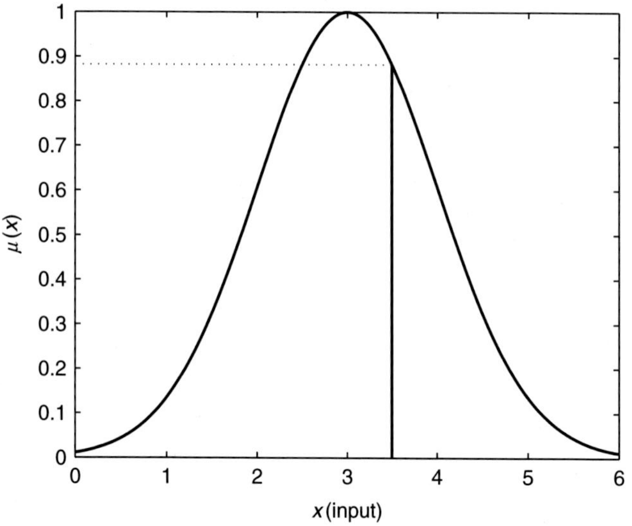

Figure 2.3 Gaussian type-1 fuzzy membership function.

2.3 BASICS OF FUZZY LOGIC CONTROL

The main idea of FLC is very well explained by Kickert and Mamdani as:

> *"The basic idea behind this approach was to incorporate the "experience" of a human process operator in the design of controller. From the set of linguistic rules which describe the operator's control strategy a control algorithm is constructed where the words are defined as fuzzy sets. The main advantage of this approach seem to be the possibility of implementing rule of thumb experience, intuition, heuristics and the fact that it does not need a model of the process" [4].*

2.3.1 FLC Block Diagram

The input of a FLC is always crisp that is fuzzified in a *fuzzification process* based on the rules in the *rule base*. After the fuzzy decisions are made by the *inference*, the output of the FLC is converted into a crisp value. This is called as *defuzzification* process.

A FLC can be divided into four main sub-groups: fuzzification, inference, rule base, and defuzzification (as shown in Fig. 2.4).

2.3.1.1 Fuzzification

Fuzzification is the process of converting a crisp input value to a fuzzy value that is performed by the use of the information in the knowledge base. Although various types of curves can be seen in literature, Gaussian,

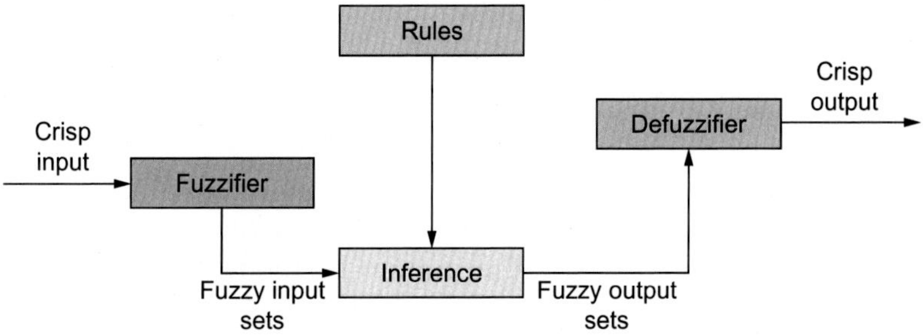

Figure 2.4 Fuzzy controller block diagram.

triangular, and trapezoidal MFs are the most commonly used in the fuzzification process. These types of MFs can easily be implemented by embedded controllers.

The MFs are defined mathematically with several parameters. In order to fine-tune the performance of a FLC, these parameters, or the shape of the MFs, can be adapted.

2.3.1.2 Rule Base

In this step, the expert knowledge is formulated as a finite number of rules. The rule base contains the rules that are to be used in making decisions. These rules are generally based on personal experience and intuition. However, in some cases, the rules can be obtained by using neural networks, genetic algorithms, or some empirical approaches.

A rule is composed of two main parts: an antecedent block (between the If and Then) and a consequent block (following Then).

If (antecedent) Then (consequent)

Although the antecedent and the consequent parts have single arguments in the above, a rule can be written with multiple arguments. While single arguments are used in SISO systems, multiple arguments are used to deal with MIMO or MISO systems.

2.3.1.3 Inference

Fuzzy decisions are produced in this process using the rules in the rule base. During this process, each rule is evaluated separately and then a decision is made for each individual rule. The result is a set of fuzzy decisions. Logical operators, such as "AND," "OR," and "NOT" define how the fuzzy variables are combined.

2.3.1.4 Defuzzification

The final step is a defuzzification process where the fuzzy output is translated into a single crisp value, like the fuzzification process, by the degree of membership values. Defuzzification is an inverse transformation compared with the fuzzification process, because in this process, the fuzzy output is converted into crisp values to be applied to the system.

There are several heuristic defuzzification methods. For instance, some methods produce an integral output considering all the elements of the resulting fuzzy set with the corresponding weights. One of the widely used methods is the center-of-area method that takes the center of gravity of the fuzzy set.

2.4 PROS AND CONS OF FUZZY LOGIC CONTROL

The classical control theory uses a mathematical model to define the relationships between the input and output of a system. The most common type of these controllers are PID controllers. After they take the output of the system and compare it with the desired input, they generate an appropriate control signal based on the calculated error value. The most serious disadvantage of these controllers is that PID controllers usually assume the system to be linear or at least it behaves as a linear system in some range. If an accurate mathematical model of a control system is available, a conventional PID controller can make the performance of the system quite acceptable. Scientists have been working on classical control theory for a long time. Today, the design of a PID type controller is very well-known subject, and its implementation is simple and cheap.

There are, of course, reasons why fuzzy logic has been famous in the last several decades. In real life, an accurate mathematical model of a control process is generally not available, and it may not even exist. The real world is nonlinear, uncertain, and always contains incomplete data. If the mathematical model is not known by the designer, there is no way to come up with a good PID controller design. The trial-and-error method may be time consuming in a complex application, especially when there exist many subsystems interacting each other. Assuming that the mathematical model is known and relatively accurate, the parameters of the system are likely to change by some outside factors, like heat, pressure, etc. In such cases, FLCs have the ability to control a system with just some limited expert knowledge. Further, FLCs are low-cost implementations based on cheap sensors.

Although fuzzy control fills an important gap in controller design methodologies that require a full mathematical clarity about a system, it has also some serious drawbacks. First of all, because fuzzy control is a method of nonlinear variable structure control, deriving its analytical structure is the first step for analytical study. However, this step is very difficult and sometimes impossible.

The second disadvantage of using a FLC is the number of design parameters. Although a classical PID controller has only three design parameters, the number of parameters for a FLC can be very large. The number and shape of input and output fuzzy sets, scaling factors, and fuzzy AND and OR operator characteristics must be determined by the designer. Moreover, there are no clear relationships between these parameters and the controller's performance.

The last disadvantage of using a FLC is that even if there exist some studies for showing the stability of the overall system [5–7], there does not exist a simple and systematic way of analyzing the stability of the system. On the other hand, this work is a very straightforward task with linear controllers.

2.5 WESTERN AND EASTERN PERSPECTIVES ON FUZZY LOGIC

It is a well-known fact that fuzzy logic has always been more popular in Eastern countries when compared to western countries. Why?

According to Buddha, almost everything contains some of its opposite. In other words, even if one belongs to one set significantly, it also belongs to its opposite set with another membership value. For instance, there is not a fully *good* or *bad* person in this world. Instead, everyone belongs to the set of *good* and *bad* with some membership values. This idea can be regarded as the basics of fuzzy logic theory where every input belongs to a set with some specific membership values.

Another philosopher, Aristotle, saw things differently. According to his logic, something could be either true or false, which is the principle foundation of mathematics and Boolean logic. This approach claims that something belongs to one set or its opposite. In such logic, the idea of being true and false simultaneously cannot exist.

The Town Hall (Dutch: Stadhuis) of Leuven in Fig. 2.5(a) is a glamorous masterpiece in Europe and a landmark building on that city's hearth. This monument was built in a Brabantine Late Gothic style between 1448 and 1469. The most significant details on this building are pointed Gothic windows, octagonal turrets on the roof, and eye-catching sculptures. On the other hand, the building in Fig. 2.5(b) is Kofukuji Buddhist temple in the city of Nara, Japan. Kofukuji was established in 669 by Kagami-no-Okimi,

Figure 2.5 Town Hall of Leuven, Belgium (a); Kofukuji Temple, Nara, Japan (b).

the wife of Fujiwara no Kamatari, wishing for her husband's recovery from illness. In Fig. 2.5(b), instead of using sharp lines, smooth and round lines were preferred, especially on the roofs.

When we consider the discussion above, it is not a big surprise that fuzzy logic theory was proposed for the first time by an Asian professor, Lotfali Askar Zadeh. Using Web of Science Core Collection, another search is made. In the first search (see Fig. 2.6(a)), the topic of "fuzzy logic control" is performed and the results are analyzed w.r.t. countries and territories.

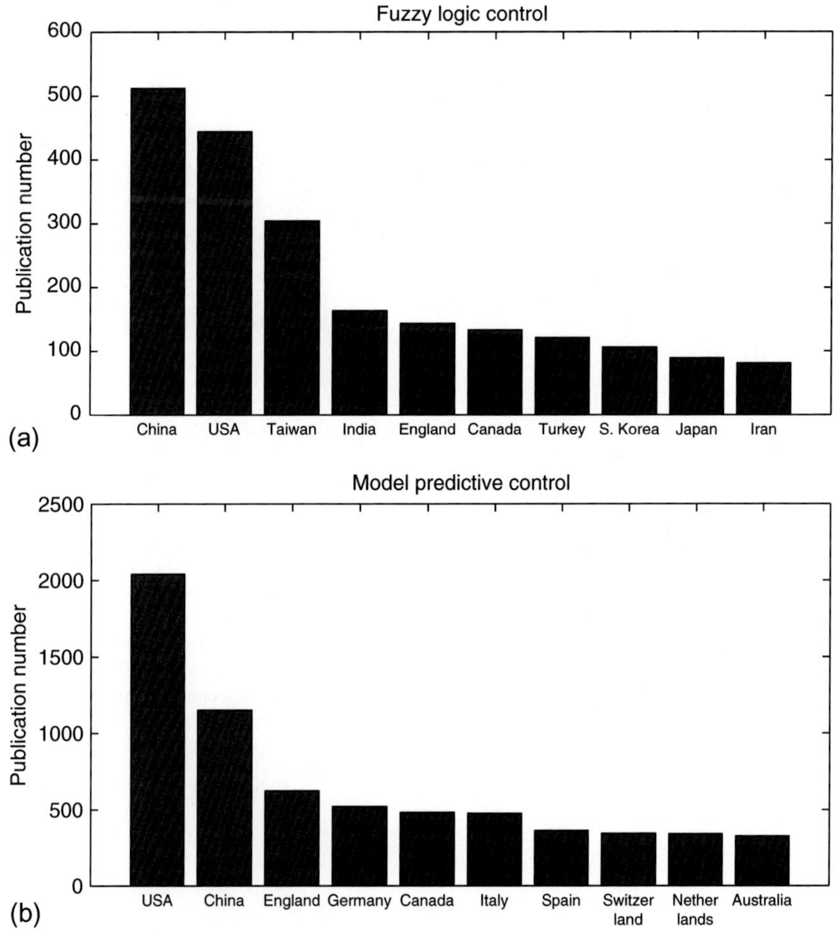

Figure 2.6 Web of Science search for "fuzzy logic control" (a); Web of Science search for "model predictive control" (b). (Accessed in June 2015).

A similar search is performed for the second time (see Fig. 2.6(b)), but for the topic of "model predictive control." Whereas the former is a model-free and heuristic approach, the latter is a model-based one. Considering the fact that there are many Asians in western countries (and vice versa), there is still a clear pattern in Fig. 2.6 shows that fuzzy logic control, as a model-free method, is more popular in Asian countries.

2.6 CONCLUSION

Real-time industrial systems require considerable man power/effort for obtaining their accurate mathematical models. Even when the model is sufficiently accurate, there are many other uncertainties, e.g., due to the precision of the sensors, noise produced by the sensors, environmental conditions of the sensors, and nonlinear characteristics of the actuators. Then, not only does the performance of the model-based approaches drastically decrease, but also the complexity of the controller design increases. In such cases, model-free approaches are generally preferred both for modeling and control purposes. The most common model-free approach is the use of fuzzy logic theory.

REFERENCES

[1] L. Zadeh, Fuzzy sets, Informat. Control 8 (1965) 338-353.
[2] L. Zadeh, Toward extended fuzzy logic—a first step, Fuzzy Sets Syst. 160 (2009) 3175-3181.
[3] E.H. Mamdani, Applications of fuzzy algorithms for control of a simple dynamic plant, Proc. IEEE 121 (1974) 1585-1588.
[4] W. Kickert, E. Mamdani, Analysis of a fuzzy logic controller, Fuzzy Sets Syst. 1 (1978) 29-44.
[5] K. Tanaka, M. Sugeno, Stability analysis and design of fuzzy control systems, Fuzzy Sets Syst. 45 (2) (1992) 135-156.
[6] Y.-Y. Cao, P. Frank, Stability analysis and synthesis of nonlinear time-delay systems via linear TakagiSugeno fuzzy models, Fuzzy Sets Syst. 124 (2) (2001) 213-229.
[7] S.G. Cao, N.W. Rees, G. Feng, Stability analysis and design for a class of continuous-time fuzzy control systems, Int. J. Control 64 (6) (1996) 1069-1087.

CHAPTER 3

Fundamentals of Type-2 Fuzzy Logic Theory

Contents

Abstract

There are two different approaches to FLS design: T1FLSs and T2FLSs. The latter is proposed as an extension of the former with the intention of being able to model the uncertainties that invariably exist in the rule base of the system. In type-1 fuzzy sets, MFs are totally certain, whereas in type-2 fuzzy sets, MFs are themselves fuzzy. The latter results in a case where the antecedents and consequents of the rules are uncertain. In this chapter, we will introduce the basics of type-2 fuzzy sets, type-2 fuzzy MFs and T2FLCs.

Keywords

Type-2 fuzzy logic theory, Type-2 fuzzy logic control, Footprint of uncertainty, Elliptic membership function, Type-2 fuzzy neural networks

3.1 INTRODUCTION

Type-2 fuzzy sets were introduced by Lotfi A. Zadeh in 1975 as an extension of type-1 fuzzy sets. Mendel and Karnik developed the theory of type-2 fuzzy sets further in their 1999 IEEE Trans. Fuzzy Syst. article [1].

http://dx.doi.org/10.1016/B978-0-12-802687-8.00003-7

The basics of the interval type-2 fuzzy systems and their design principles are described in Ref. [2].

T2FLSs appear to be a more promising method than their type-1 counterparts for handling uncertainties such as noisy data and changing environments [3, 4]. In Refs. [5, 6] the effects of the measurement noise in T1FLCs and T2FLCs and identifiers are simulated to perform a comparative analysis. It was concluded that the use of T2FLCs in real-world applications [7], which exhibit measurement noise and modeling uncertainties, can be a better option than T1FLCs.

There are (at least) eight sources of uncertainties in FLSs:

1. Precision of the measurement devices
2. Noise on the measurement devices
3. Environmental conditions of the measurement devices
4. Unknown nonlinear characteristics of the actuators
5. Lack of modeling
6. The meanings of the words that are used in the antecedents and consequents of rules can be uncertain (words mean different things to different people) [8]
7. Consequents may have a histogram of values associated with them, especially when knowledge is extracted from a group of experts who do not all agree [8]
8. Uncertainty caused by some unvisited data that the fuzzy system does not have any predefined rules for

When a system has large amount of uncertainties, T1FLSs may not be able to achieve the desired level of performance with reasonable complexity of the structure [9]. In such cases, the use of T2FLSs is suggested as the preferable approach in literature in many areas, such as forecasting of time-series [10], controlling of mobile robots [11], and the truck backing-up control problem [12]. The VLSI and FPGA implementations of T2FLSs are also discussed in Refs. [13, 14]. In Ref. [10], it is shown that when the parameters are tuned appropriately, T2FLSs can result in a better prediction ability as compared to T1FLSs. In Ref. [11], T2FLSs are applied to real-time mobile robots for indoor and outdoor environments. The real-time implementation studies show that a traditional T1FLC cannot handle high levels of uncertainties in the system effectively and a T2FLC using type-2 fuzzy sets results in better performance. Moreover, with the latter approach, the number of rules to be determined may be reduced (it should be noted that this may not mean a corresponding decrease in the parameters to be updated). In Ref. [12], the authors construct an interval T2FNN structure, and show that a better performance can be obtained as compared to a conventional T1FNN. The VLSI implementation of T2FLSs is also discussed in literature and it has been shown that the inference speed can be sufficiently high for real-time applications. In Ref. [14], a type-2 self-organizing neural fuzzy system and its hardware implementation is proposed. It is reported that using interval type-2 fuzzy sets in that structure enables the overall system to be more robust than the one with type-1 fuzzy systems.

There exist two main approaches to the design of a FLC:

1. Type-1 fuzzy sets: MFs are totally certain.
2. Type-2 fuzzy sets: MFs that are themselves fuzzy. This results in antecedents and consequents of the rules being uncertain.

3.2 TYPE-2 FUZZY SETS

A type-2 fuzzy set, $\tilde{A}$, may be represented as [8]:

$$\tilde{A} = \{((x, u), \mu_{\tilde{A}}(x, u))| \quad \forall x \in X \quad \forall u \in J_x \subseteq [0, 1]\} \tag{3.1}$$

where $\mu_{\tilde{A}}(x, u)$ is the type-2 fuzzy MF in which $0 \leq \mu_{\tilde{A}}(x, u)) \leq 1$. $\tilde{A}$ can also be defined as [8]:

$$\tilde{A} = \int_{x \in X} \int_{u \in J_x} \mu_{\tilde{A}}(x, u)/(x, u) \quad J_x \subseteq [0, 1] \tag{3.2}$$

where $\int \int$ denotes the union over all admissible x and u [8].

J_x is called primary membership of x [8]. Additionally, there is a *secondary membership* value corresponding to each primary membership value that defines the possibility for primary memberships [15]. While the secondary MFs can take values in the interval of [0,1] in generalized T2FLSs, they are uniform functions that only take on values of 1 in interval T2FLSs. Since the general T2FLSs are computationally demanding, the use of interval T2FLSs is more commonly seen in the literature, due to the fact that the computations are more manageable.

If the circumstances are so fuzzy, the places of the MFs may not be determined precisely. In such cases, the membership grade cannot be determined as a crisp number in [0,1], and the use of type-2 fuzzy sets might be a preferable option.

If the standard deviation of a type-1 fuzzy Gaussian MF is blurred, Fig. 3.1 can be obtained. In Fig. 3.1, the MF does not have a single

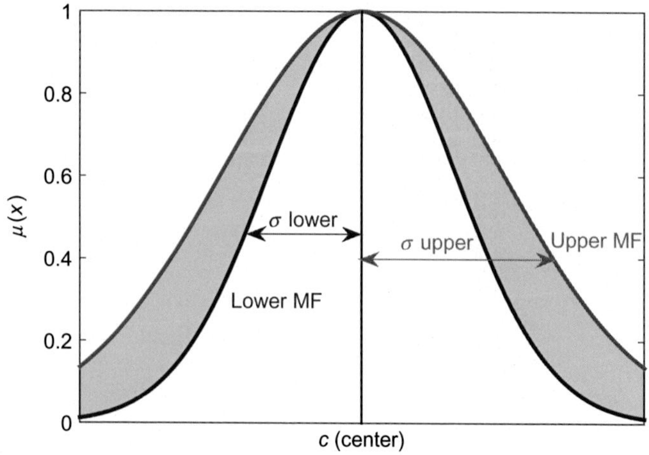

Figure 3.1 Gaussian type-2 fuzzy MF (FOU).

value for a specific value of an input. The values for a specific input, the vertical line intersecting the MFs, do not need to all be weighted the same. Moreover, an amplitude distribution can be assigned to all of those points. Hence, a three-dimensional MF-a type-2 MF- that characterizes a type-2 fuzzy set is created if the amplitude distribution operation is done for all inputs [9].

The FOU, the union of all primary memberships, is said to be the bounded region that represents the uncertainty in the primary memberships of a type-2 fuzzy set. In Fig. 3.1, an upper MF (red) and a lower MF (blue) are two type-1 MFs that are the bounds for the FOU of a type-2 fuzzy set [9]. It is assumed to have infinite type-1 MFs between the lower and upper MFs in a type-2 fuzzy set.

3.2.1 Interval Type-2 Fuzzy Sets

When all $\mu_{\tilde{A}}(x, u)$ are equal to 1, then $\tilde{A}$ is an interval type-2 fuzzy set. The special case of (3.2) might be defined for the interval type-2 fuzzy sets:

$$\tilde{A} = \int_{x \in X} \int_{u \in J_x} 1/(x, u) \quad J_x \subseteq [0, 1] \tag{3.3}$$

Researchers are familiar with the computational burden of general T2FLS. Hence, interval T2FLSs are commonly used in literature. Both the general and interval type-2 fuzzy MFs are three-dimensional. As can be seen from Fig. 3.2, the only difference between them is that the secondary

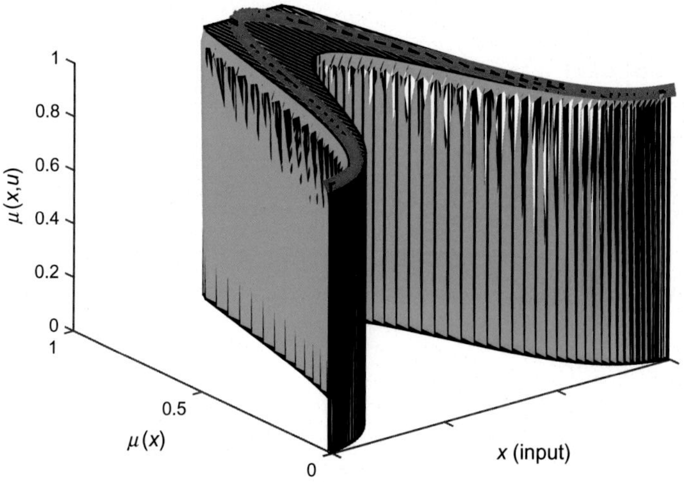

Figure 3.2 Three-dimensional representation of interval type-2 fuzzy MFs.

membership value of the interval type-2 MF is equal to 1. In this book, it is focused on the interval T2FLSs.

3.2.2 T2FLS Block Diagram

The T2FLS block diagram is shown in Fig. 3.3. The reader should have basic knowledge of T1FLSs, since only the similarities and the differences between T1FLSs and T2FLSs will be given in this book. As can be seen from Fig. 3.3, an additional block to Fig. 2.4 (type reduction) is needed in T2FLS design. Although the structure in Fig. 3.3 brings some advantages when dealing with uncertainties, it also increases the computational burden.

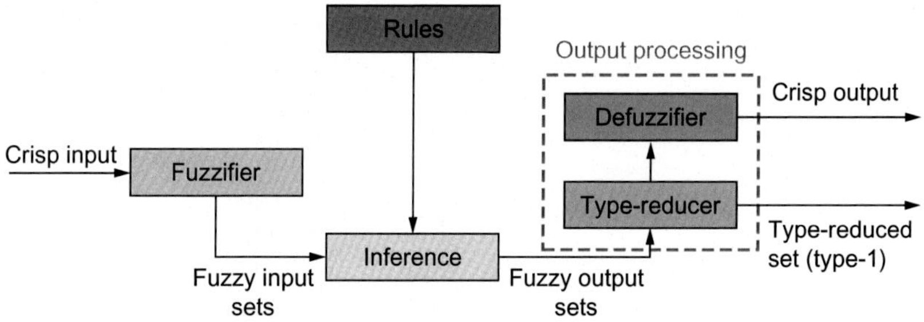

Figure 3.3 T2FLS block diagram.

The following are the basic blocks of a T2FLS:

3.2.2.1 Fuzzifier

The fuzzifier maps crisp inputs into type-2 fuzzy sets, which activates the inference engine.

3.2.2.2 Rule Base

The rules in a T2FLS remain the same as in a T1FLS, but antecedents and consequents are represented by interval type-2 fuzzy sets.

3.2.2.3 Inference

The inference block assigns fuzzy inputs to fuzzy outputs using the rules in the rule base and operators such as union and intersection. In type-2 fuzzy sets, *join* ($\sqcup$) and *meet* operators ($\sqcap$), which are new concepts in fuzzy logic theory, are used instead of *union* and *intersection* operators. These two new operators are used in secondary MFs, and they are defined and explained in detail in Ref. [16].

3.2.2.4 Type Reduction

The type-2 fuzzy outputs of the inference engine are transformed into type-1 fuzzy sets that are called *the type-reduced sets*. There are two common methods for the type-reduction operation in the interval T2FLSs: One is the Karnik-Mendel iteration algorithm, and the other is Wu-Mendel uncertainty bounds method. These two methods are based on the calculation of the centroid.

3.2.2.5 Defuzzification

The outputs of the type reduction block are given to the defuzzificaton block. The type-reduced sets are determined by their left end point and right end point, and the defuzzified value is calculated by the average of these points.

3.3 EXISTING TYPE-2 MEMBERSHIP FUNCTIONS

There exists number of type-2 fuzzy MFs in literature, i.e., triangular, Gaussian, trapezoidal, sigmoidal, pi-shaped, etc. Gaussian type MFs are widely used in which uncertainties can be associated with mean and standard deviation of the Gaussian function.

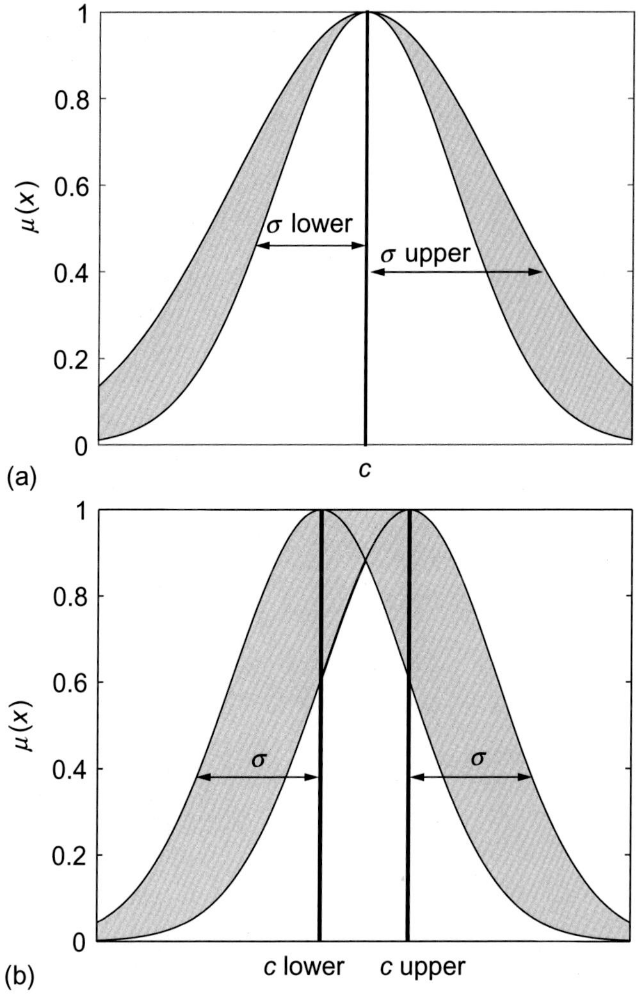

Figure 3.4 Type-2 MFs with uncertain standard deviation (a) and uncertain mean (b).

In Fig. 3.4(a) and (b), Gaussian type-2 fuzzy MFs with uncertain standard deviation and uncertain mean are shown. The MF is expressed as:

$$\tilde{\mu}(x) = \exp\left(-\frac{1}{2}\frac{(x-c)^2}{\sigma^2}\right) \tag{3.4}$$

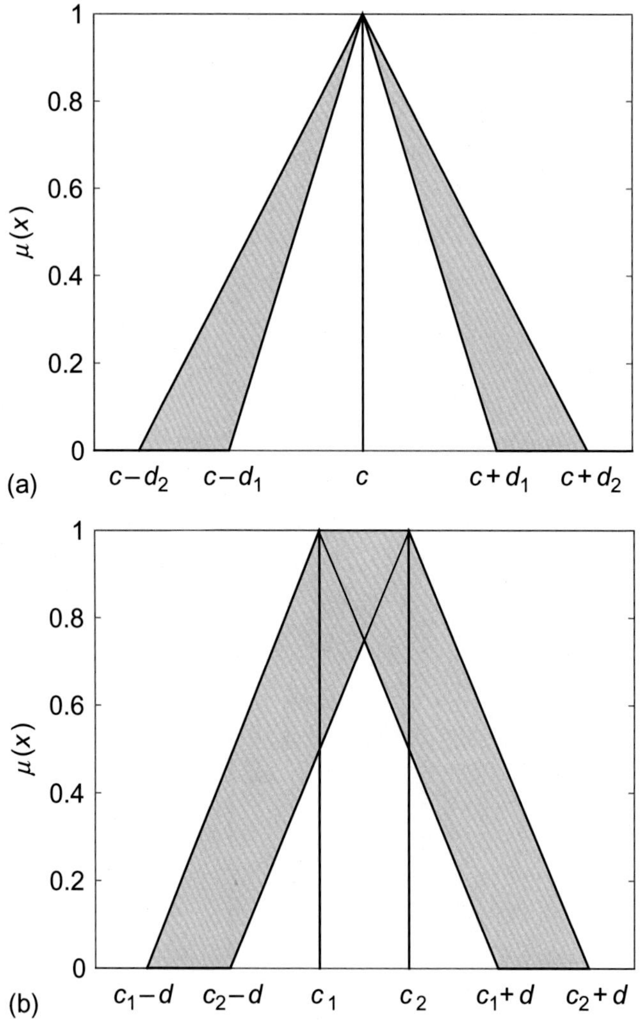

Figure 3.5 Type-2 MFs with uncertain width (a) and center (b).

where c and σ are the center and the width of the MF and x is the input vector.

In Fig. 3.5(a) and (b), triangular type-2 fuzzy sets with uncertain width and uncertain center are shown. The MF is expressed as:

$$\tilde{\mu}(x) = \begin{cases} 1 - \frac{|x-c|}{d} & \text{if } c-d < x < c+d \\ 0 & \text{else} \end{cases} \tag{3.5}$$

where c and d are the center and the width of the MF and x is the input vector.

3.3.1 A Novel Type-2 MF: Elliptic MF

A novel type-2 fuzzy MF is introduced in this section. It has certain values on both ends of the support and the kernel, and some uncertain values on the other values of the support. The need for proposing such a novel MF will be explained in detail in Chapter 9. The lower ($\underline{\mu}$) and the upper ($\bar{\mu}$) MFs with the parameters c, d, a_1, and a_2 are defined as follows:

$$\bar{\mu}(x) = \begin{cases} \left(1 - |\frac{x-c}{d}|^{a_1}\right)^{1/a_1} & \text{if } c-d < x < c+d \\ 0 & \text{else} \end{cases} \tag{3.6}$$

$$\underline{\mu}(x) = \begin{cases} \left(1 - |\frac{x-c}{d}|^{a_2}\right)^{1/a_2} & \text{if } c-d < x < c+d \\ 0 & \text{else} \end{cases} \tag{3.7}$$

where c and d are the center and the width of the MF and x is the input vector. The parameters a_1 and a_2 determine the width of the uncertainty of the proposed MF, and these parameters should be selected as follows:

$$\begin{aligned} a_1 &> 1 \\ 0 < a_2 &< 1 \end{aligned} \tag{3.8}$$

Figure 3.6(a), (b), and (c) show the shapes of the proposed MF for $a_1 = a_2 = 1$, $a_1 = 1.2, a_2 = 0.8$, and $a_1 = 1.4, a_2 = 0.6$, respectively. As can be seen from Fig. 3.6a, the shape of the proposed type-2 MF is changed to a type-1 triangular MF when its parameters are selected as $a_1 = a_2 = 1$. These parameters can be selected as some constants or they can be tuned adaptively.

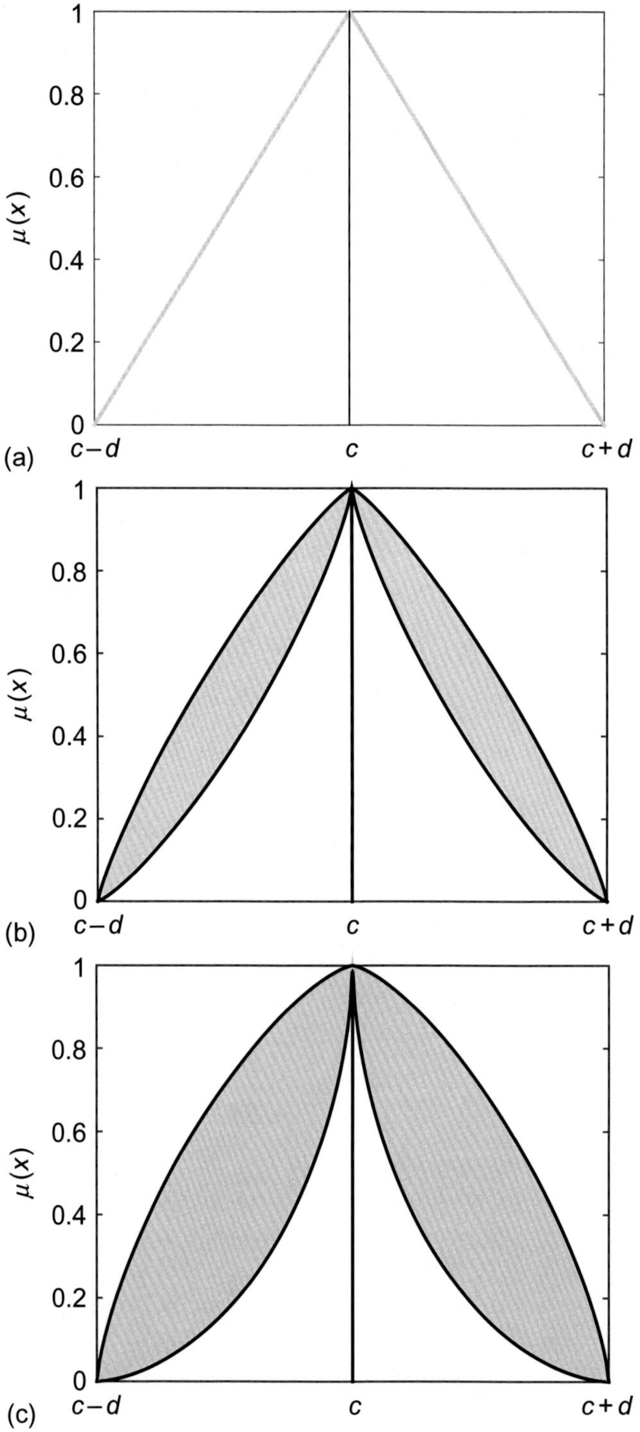

Figure 3.6 Shapes of the proposed type-2 MF with different values for a_1 and a_2.

3.4 CONCLUSION

If the circumstances are so fuzzy, the places of the type-1 MFs may not be determined precisely. In cases where the membership grade cannot be determined as a crisp number in [0,1], the use of type-2 fuzzy sets might be a preferable option. T2FLSs appear to be a more promising method than their type-1 counterparts for handling uncertainties such as noisy data and changing working environments.

REFERENCES

[1] N.N. Karnik, J.M. Mendel, Q. Liang, Type-2 fuzzy logic systems, IEEE Trans. Fuzzy Syst. 7 (1999) 643-658.

[2] J.M. Mendel, R. John, F. Liu, Interval Type-2 fuzzy logic systems made simple, IEEE Trans. Fuzzy Syst. 14 (2006) 808-821.

[3] C.-F. Juang, C.-H. Hsu, Reinforcement interval Type-2 fuzzy controller design by online rule generation and Q-value-aided ant colony optimization, IEEE Trans. Syst. Man Cybernet. B Cybernet. 39 (6) (2009) 1528-1542.

[4] C. Juang, Y. Tsao, A self-evolving interval Type-2 fuzzy neural network with online structure and parameter learning, IEEE Trans. Fuzzy Syst. 16 (2008) 1411-1424.

[5] R. Sepulveda, P. Melin, A. Rodriguez, A. Mancilla, O. Montiel, Analyzing the effects of the footprint of uncertainty in type-2 fuzzy logic controllers, Eng. Lett. 13 (2006) 138-147.

[6] M.A. Khanesar, M. Teshnehlab, E. Kayacan, O. Kaynak, A novel Type-2 fuzzy membership function: application to the prediction of noisy data, in: Proceedings of the IEEE International Conference on Computational Intelligence for Measurement System and Applications, 2010, pp. 128-133.

[7] M. Biglarbegian, W. Melek, J. Mendel, Design of novel interval Type-2 fuzzy controllers for modular and reconfigurable robots: theory and experiments, IEEE Trans. Indust. Electron. 58 (4) (2011) 1371-1384.

[8] J.M. Mendel, R.I.B. John, Type-2 fuzzy sets made simple, IEEE Trans. Fuzzy Syst. 10 (2002) 117-127.

[9] J.M. Mendel, Uncertain Rule-Based Fuzzy Logic System: Introduction and New Directions, Prentice Hall, Upper Saddle River, NJ, 2001.

[10] N. Karnik, J. Mendel, Applications of type-2 fuzzy logic systems to forecasting of time-series, Informat. Sci. 120 (1999) 89-111.

[11] H.A. Hagras, A hierarchical type-2 fuzzy logic control architecture for autonomous mobile robots, IEEE Trans. Fuzzy Syst. 12 (4) (2004) 524-539.

[12] C. Wang, C. Cheng, T. Lee, Dynamical optimal training for interval type-2 fuzzy neural network (T2FNN), IEEE Trans. Syst. Man Cybernet. B Cybernet. 34 (2004) 1462-1477.

[13] S. Huang, Y. Chen, VLSI implementation of type-2 fuzzy inference processor, in: Proceedings of IEEE ISCAS 2005, 2005, pp. 3307-3310.

[14] C. Juang, Y. Tsao, A type-2 self-organizing neural fuzzy system and its FPGA implementation, IEEE Trans. Syst. Man Cybernet. B Cybernet. 38 (6) (2008) 1537-1548.

[15] Q. Liang, J.M.Mendel, Interval type-2 fuzzy logic systems: theory and design, IEEE Trans. Fuzzy Syst. 8 (2000) 535-550.

[16] N.N. Karnik, J.M. Mendel, Operations on type-2 fuzzy sets, Fuzzy Sets Syst. 122 (2) (2001) 327-348.

CHAPTER 4

Type-2 Fuzzy Neural Networks

Contents

Abstract

The two most common artificial intelligence techniques, FLSs and ANNs, can be used in the same structure simultaneously, namely as "fuzzy neural networks." The advantages of ANNs such as learning capability from input-output data, generalization capability, and robustness and the advantages of fuzzy logic theory such as using expert knowledge are harmonized in FNNs. In this chapter, type-1 and type-2 TSK fuzzy logic models are introduced. Instead of using fuzzy sets in the consequent part (as in Mamdani models), the TSK model uses a function of the input variables. The order of the function determines the order of the model, e.g., zeroth-order TSK model, first-order TSK model, etc.

Keywords

Type-1 fuzzy neural networks, Type-2 fuzzy neural networks, TSK models, Artificial intelligence, Fuzzy logic, Neural networks

4.1 TYPE-1 TAKAGI-SUGENO-KANG MODEL

A type-1 TSK model can be described by fuzzy If-Then rules. For instance, in a first-order type-1 TSK model, the rule base is as follows:

$$\text{IF } x_1 \text{ is } A_{j1} \text{ and } x_2 \text{ is } A_{j2} \text{ and } \ldots \text{and } x_n \text{ is } A_{jn} \qquad (4.1)$$
$$\text{THEN } u_j = \sum_{i=1}^{n} w_{ij} x_i + b_j$$

Fuzzy Neural Networks for Real Time Control Applications
http://dx.doi.org/10.1016/B978-0-12-802687-8.00004-9

where $x_1, x_2, \ldots, x_n$ are the input variables, u_j's are the output variables, and A_{ji}'s are type-1 fuzzy sets for the jth rule and the ith input. The parameters in the consequent part of the rules are w_{ij} and b_j $(i = 1, \ldots, n, j = 1, \ldots, M)$.

The final output of the system can be written as:

$$u = \frac{\sum_{j=1}^{M} f_j u_j}{\sum_{j=1}^{M} f_j} \tag{4.2}$$

where f_j is given by:

$$f_j(x) = \mu_{A_{j1}}(x_1) * \cdots * \mu_{A_{jn}}(x_n) \tag{4.3}$$

in which $*$ represents the t-norm, which is the *prod* operator in this book.

4.2 OTHER TAKAGI-SUGENO-KANG MODELS

Other TSK models (shown in Table 4.1) can be classified into three groups [1]:

1. Model I: Antecedents are type-2 fuzzy sets, and consequents are type-1 fuzzy sets (A2-C1)
2. Model II: Antecedents are type-2 fuzzy sets, and consequents are crisp numbers (A2-C0)
3. Model III: Antecedents are type-1 fuzzy sets, and consequents are type-1 fuzzy sets (A1-C1)

4.2.1 Model I

Type-2 Model I can be described by fuzzy If-Then rules in which the antecedent part is type-2 fuzzy sets. In the consequent part, the structure is similar to that of a type-1 TSK fuzzy system, however, the parameters are type-1 fuzzy sets rather than crisp numbers. They are therefore named as "Type-2 TSK Model I" systems. In Model I, the rule base is as follows:

Table 4.1 Classification of other TSK models

Other TSK FLSs	Model I	Model II	Model III
Antecedent	Type-2 fuzzy sets	Type-2 fuzzy sets	Type-1 fuzzy sets
Consequent	Type-1 fuzzy sets	Crisp numbers	Type-1 fuzzy sets

$$\text{IF } x_1 \text{ is } \tilde{A}_{j1} \text{ and } x_2 \text{ is } \tilde{A}_{j2} \text{ and } \ldots \text{and } x_n \text{ is } \tilde{A}_{jn} \qquad (4.4)$$

$$\text{THEN } U_j = \sum_{i=1}^{n} W_{ij}x_i + B_j$$

where $x_1, x_2, \ldots, x_n$ are the input variables, U_j's are the output variables, and $\tilde{A}_{ji}$'s are type-2 fuzzy sets for the jth rule and the ith input. The parameters in the consequent part of the rules are W_{ij} and B_j $(i = 1, \ldots, n, j = 1, \ldots, M)$, which are type-1 fuzzy sets. The final output of the first-order type-2 TSK Model I is as follows [1]:

$$U(U_1, \ldots, U_M, F_1, \ldots, F_M) = \int_{u_1} \cdots \int_{u_M} \int_{f_1} \cdots \int_{f_M} \tau_{j=1}^{M} \mu_{U_j}(u_j) \star \tau_{j=1}^{M} \mu_{F_j}(f_j) \Bigg/ \frac{\sum_{j=1}^{M} f_j u_j}{\sum_{j=1}^{M} f_j} \qquad (4.5)$$

where M is the number of rules fired, $u_j \in U_j, f_j \in F_j$, and τ and $\star$ indicate the t-norm. F_j is the firing strength which is defined as:

$$F_j = \mu_{\tilde{A}_{j1}}(x_1) \sqcap \mu_{\tilde{A}_{j2}}(x_2) \sqcap \cdots \sqcap \mu_{\tilde{A}_{jn}}(x_n) \qquad (4.6)$$

where $\sqcap$ shows the meet operation.

Although the calculation of (4.5) is difficult, some general concepts are explained in Ref. [2]. When interval type-2 sets are used in the antecedent part and type-1 sets are used in the consequent part, it is shown in Ref. [1] that the output of an interval T2FLS is:

$$f(x) = \frac{u_l + u_r}{2} \qquad (4.7)$$

where u_r and u_l are the maximum and minimum values of u, respectively.

The reader is encouraged to refer [1] and [2] for further information about the calculation process of u_r and u_l. A broad survey exists in literature in which number of different type reducers for Model I are compared [4].

4.2.2 Model II

Model II can be regarded as a special case of Model I where the antecedents are type-2 fuzzy sets and the consequents are polynomials. A type-2 TSK

Model II can be described by fuzzy If-Then rules. For instance, in a first-order type-2 TSK Model II, the rule base is as follows [1]:

$$\text{IF } x_1 \text{ is } \tilde{A}_{j1} \text{ and } x_2 \text{ is } \tilde{A}_{j2} \text{ and } \ldots \text{and } x_n \text{ is } \tilde{A}_{jn} \qquad (4.8)$$
$$\text{THEN } u_j = \sum_{i=1}^{n} w_{ij}x_i + b_j$$

where $x_1, x_2, \ldots, x_n$ are the input variables, u_j's are the output variables, $\tilde{A}_{ji}$'s are type-2 fuzzy sets for the jth rule and the ith input. The parameters in the consequent part of the rules are w_{ij} and b_j $(i = 1, \ldots, n, j = 1, \ldots, M)$.

The final output of the model is as follows [1]:

$$U(F_1, \ldots, F_M) = \int_{f_1} \cdots \int_{f_M} \tau_{j=1}^{M} \mu_{F_j}(f_j) \Bigg/ \frac{\sum_{j=1}^{M} f_j u_j}{\sum_{j=1}^{M} f_j} \qquad (4.9)$$

where M is the number of rules fired, $f_j \in F_j$, and τ indicates the t-norm.

Note that (4.9) is a special case of (4.5), because each U_j in (4.5) is converted into a crisp value here. The firing strength is the same as (4.6).

4.2.2.1 Interval Type-2 TSK FLS

In the structure of the interval type-2 TSK FLS, (4.9) is given as follows [3]:

$$Y_{TSK/A2-C0} = \int_{f^1 \in [\underline{f}^1, \overline{f}^1]} \cdots \int_{f^M \in [\underline{f}^M, \overline{f}^M]} 1 \Bigg/ \frac{\sum_{j=1}^{M} f_j u_j}{\sum_{j=1}^{M} f_j} \qquad (4.10)$$

where $\underline{f}_j$ and $\overline{f}_j$ are given by:

$$\underline{f}_j(x) = \underline{\mu}_{A_{j1}}(x_1) * \cdots * \underline{\mu}_{A_{jn}}(x_n) \qquad (4.11)$$
$$\overline{f}_j(x) = \overline{\mu}_{A_{j1}}(x_1) * \cdots * \overline{\mu}_{A_{jn}}(x_n)$$

in which $*$ represents the t-norm, which is the *prod* operator in this book.

The output of the fuzzy system in closed form is approximated by [3]:

$$Y_{TSK/A2-C0} = \frac{\sum_{j=1}^{M} \underline{f}_j u_j}{\sum_{j=1}^{M} \underline{f}_j + \sum_{j=1}^{M} \overline{f}_j} + \frac{\sum_{j=1}^{M} \overline{f}_j u_j}{\sum_{j=1}^{M} \underline{f}_j + \sum_{j=1}^{M} \overline{f}_j} \qquad (4.12)$$

4.2.2.2 Numerical Example of the Interval Type-2 TSK FLS

In order to be able to give a clear explanation about the inference of this type FLS, a numerical example is given:

Let's assume a T2FLS (A2-CO) with two inputs and two type-2 fuzzy MFs for each. While the antecedent type-2 fuzzy MFs are given in Fig. 4.1, the consequent part of the rules are given as follows: $u_1 = 4x_1 + x_2$ and $u_2 = 2x_1 + 3x_2$.

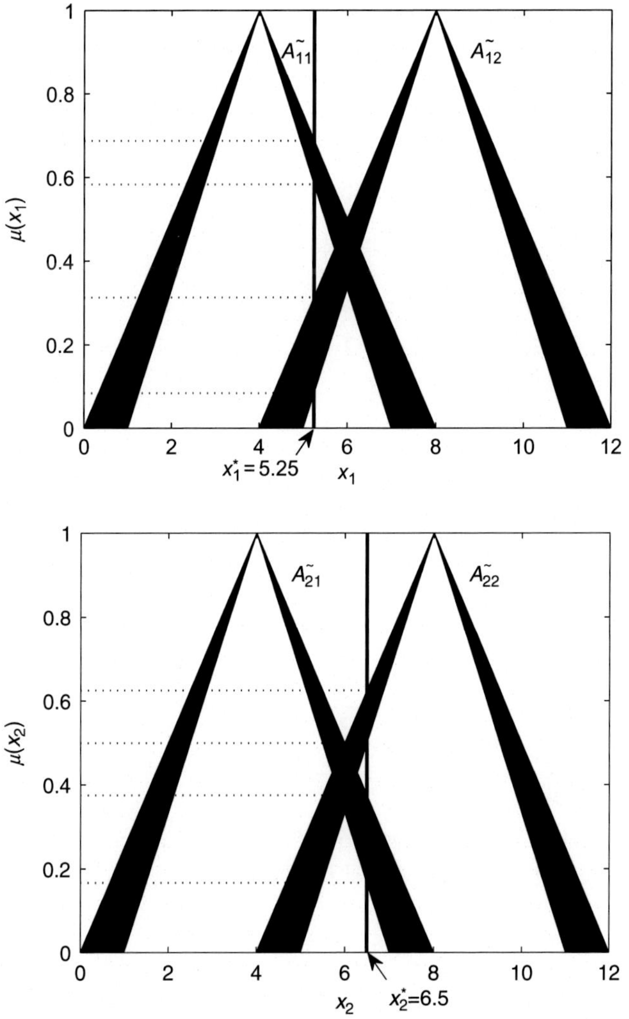

Figure 4.1 Two rules each having two type-2 triangular fuzzy MFs.

The mathematical form of triangular MFs are as follows:

$$\tilde{\mu}(x) = \begin{cases} 1 - \frac{|x-c|}{d} & \text{if } c - d < x < c + d \\ 0 & \text{else} \end{cases} \quad (4.13)$$

where $\underline{c}_{11} = \underline{c}_{21} = \bar{c}_{11} = \bar{c}_{21} = 4$, $\underline{c}_{12} = \underline{c}_{22} = \bar{c}_{12} = \bar{c}_{22} = 8$, and $\underline{d}_{11} = \underline{d}_{21} = \underline{d}_{12} = \underline{d}_{22} = 3$, $\bar{d}_{11} = \bar{d}_{21} = \bar{d}_{12} = \bar{d}_{22} = 4$.

The input 1 (x_1^*) and the input 2 (x_2^*) are selected as 5.25 and 6.5, respectively. The firing strengths are as follows:

$$\begin{aligned} \bar{f}_1 &= 0.6875 * 0.6250 = 0.4297 \\ \underline{f}_1 &= 0.5833 * 0.5000 = 0.2917 \\ \bar{f}_2 &= 0.3125 * 0.3750 = 0.1172 \\ \underline{f}_2 &= 0.0833 * 0.1667 = 0.0139 \end{aligned} \quad (4.14)$$

$$\begin{aligned} u_1 &= 4x_1 + x_2 = 4 * 5.25 + 6.5 = 27.50 \\ u_2 &= 2x_1 + 3x_2 = 2 * 5.25 + 3 * 6.5 = 30.00 \end{aligned} \quad (4.15)$$

$$\begin{aligned} u_l &= \frac{0.4297 * 27.50 + 0.0139 * 30.00}{0.4297 + 0.0139} = 27.5783 \\ u_r &= \frac{0.2917 * 27.50 + 0.1172 * 30.00}{0.2917 + 0.1172} = 28.2166 \end{aligned} \quad (4.16)$$

The final output is calculated as follows:

$$u^* = \frac{u_l + u_r}{2} = 27.8974 \quad (4.17)$$

4.2.3 Model III

The TSK Model III can be described by fuzzy If-Then rules. For instance, in a first-order type-2 TSK Model III the rule base is as follows [1]:

$$\begin{aligned} &\text{IF } x_1 \text{ is } A_{j1} \text{ and } x_2 \text{ is } A_{j2} \text{ and } \ldots \text{and } x_n \text{ is } A_{jn} \\ &\qquad \text{THEN } U_j = \sum_{i=1}^{n} W_{ij} x_i + B_j \end{aligned} \quad (4.18)$$

where $x_1, x_2, \ldots, x_n$ are the input variables, U_j's are the output variables, and A_{ji}'s are type-1 fuzzy sets for jth rule and the ith input. The parameters in the consequent part of the rules are W_{ij} and B_j $(i = 1, \ldots, n, j = 1, \ldots, M)$. Note that both the consequent parameters and the outputs of the rules above are type-1 fuzzy sets. Also, A_{jk}'s are type-1 fuzzy sets $(k = 1, \ldots, n)$.

The final output of the model is as follows:

$$U(U_1, \ldots, U_M) = \int_{u_1} \cdots \int_{u_M} \tau_{j=1}^{M} \mu_{U_j}(u_j) \frac{\sum_{j=1}^{M} f_j u_j}{\sum_{j=1}^{M} f_j} \tag{4.19}$$

where M is the number of rules fired, $u_j \in U_j$.

f_j is the firing strength, which is defined as:

$$f_j = \mu_{A_{j1}}(x_1) * \mu_{A_{j2}}(x_2) * \cdots * \mu_{A_{jn}}(x_n) \tag{4.20}$$

4.3 CONCLUSION

The advantages of ANNs such as learning capability from input-output data, generalization capability, and robustness and the advantages of fuzzy logic theory such as using expert knowledge are harmonized in FNNs. Inspired by the conventional FNNs, T2FNNs have been designed in which the MFs are type-2. These systems are stronger to deal with uncertainties in the rule base of the system compared to their type-1 counterparts.

REFERENCES

[1] Q. Liang, Fading channel equalization and video traffic classification using nonlinear signal processing techniques, University of Southern California, USA, 2000.

[2] N. Karnik, J. Mendel, An introduction to type-2 fuzzy logic systems 1998, October 1998, URL http://sipi.usc.edu/mendel/report.

[3] M. Begian, W. Melek, J. Mendel, Parametric design of stable type-2 TSK fuzzy systems, in: Fuzzy Information Processing Society, 2008. NAFIPS 2008. Annual Meeting of the North American, 2008, pp. 1-6.

[4] Wu. Dongrui, Approaches for reducing the computational cost of interval type-2 fuzzy logic systems: overview and comparisons. Fuzzy Systems, IEEE Transactions on 21, no. 1 (2013) 80-99.

CHAPTER 5

Gradient Descent Methods for Type-2 Fuzzy Neural Networks

Contents

Abstract

Given an initial point, if an algorithm tries to follow the negative of the gradient of the function at the current point to reach a local minimum, we face the most common iterative method to optimize a nonlinear function: the GD method. The main goal of this chapter is to briefly discuss a multivariate optimization technique, namely the GD algorithm, to optimize a nonlinear unconstrained problem. The referred optimization problem is a cost function of a FNN, either type-1 or type-2, in this chapter. The main features, drawbacks and stability conditions of these algorithms are discussed.

Keywords

Gradient descent for FNN, Levenberg-Marquardt for FNN, Momentum-term gradient descent, Adaptive learning rate, Gradient descent, Stability analysis

Fuzzy Neural Networks for Real Time Control Applications
http://dx.doi.org/10.1016/B978-0-12-802687-8.00005-0

5.1 INTRODUCTION

Optimization methods are ubiquitous when it comes to the estimation of the parameters of a FNN, either type-1 or type-2. In the case of using a FNN as a controller, identifier or classifier, the design process of the corresponding FNN is nonlinear, and in most cases, an unconstrained optimization problem. The first step is to define an objective function that may also be called a cost function or performance index. In controller design, the objective function is defined based on the difference between the output of the plant and the reference signal. In the design of an identifier, the cost function is defined on the basis of the difference between the output of the FNN and the measured output of the system. When one wants to design a classifier, the difference between the class number of an input with respect to its real class number may be used as the cost function.

The purpose of this chapter is to briefly discuss the iterative numerical methods based on gradient descent algorithm to optimize a nonlinear unconstraint problem in order to design a FNN. The main features, drawbacks and stability conditions of these algorithms are also considered.

5.2 OVERVIEW OF ITERATIVE GRADIENT DESCENT METHODS

Consider the following nonlinear optimization problem:

Minimize the scalar cost function $F(x)$
subject to $x \in \mathbb{R}^n$

There exist two kinds of minimums for $F(x)$: *local minimum*(see Fig. 5.1) and *global minimum* (see Fig. 5.1). The solution x^* is said to be a local minimum of the function if there exists no better point in its neighborhood or:

$$F(x^*) \leq F(x) \quad \forall x \quad \text{s.t.} \quad \|x - x^*\| < \epsilon, \; 0 < \epsilon \tag{5.1}$$

on the other hand, x^* is defined as a global minimum if there exists no better vector in all other possible vectors, i.e.,

$$F(x^*) \leq F(x), \quad \forall x \in \mathbb{R}^n \tag{5.2}$$

Although the main goal of an optimization technique is always to find the global minimum of the cost function, the local minimum exists

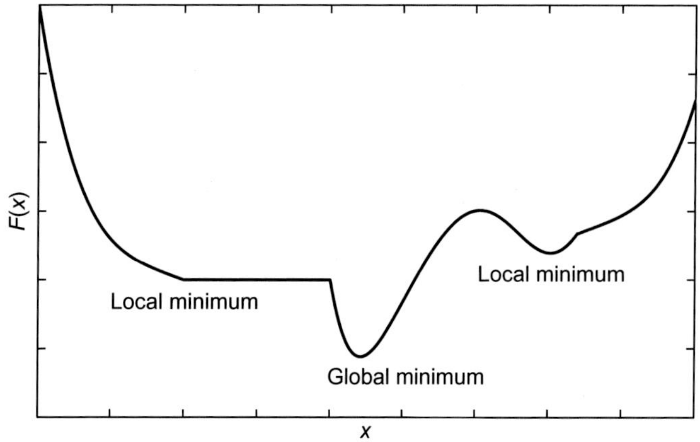

Figure 5.1 Local minimum and global minimum in one-dimensional space.

inevitably in the parameter estimation space of the FNN. Thus, it is possible that the iterative optimization algorithms may trap a local minimum.

5.2.1 Basic Gradient-Descent Optimization Algorithm

Since a cost function is a scalar function having several dimensions, the basic derivative must be generalized to cover higher dimensional functions. Such a generalization of the simple derivative to higher dimensions results in a new operator, namely the *gradient*. The symbol "∇" represents this operator.

Given a vector $x \in \mathbb{R}^n$ with $\nabla F(x) \neq 0$, the GD update rule to find the minimum is given as:

$$x(k+1) = x(k) - \alpha \nabla F(x), \quad \forall \alpha > 0 \tag{5.3}$$

The first-order Taylor expansion of $F(x(k+1))$ around $x(k)$ is given by:

$$F(x(k+1)) = F(x(k)) + (\nabla F(x(k)))^{\mathrm{T}} (x(k+1) - x(k)) + o(\|x(k+1) - x(k)\|) \tag{5.4}$$

Considering the update rule as in (5.3), we have:

$$F(x(k+1)) = F(x(k)) - \alpha (\nabla F(x(k)))^{\mathrm{T}} \nabla F(x(k)) + o(\alpha \|\nabla F(x(k))\|) \tag{5.5}$$

The term $\alpha(\nabla F(x(k)))^{\mathrm{T}}\nabla F(x(k))$ dominates the higher-order terms $o(\alpha\|\nabla F(x(k))\|)$ near zero [1]. Hence, for small variations, $F(x(k+1)) < F(x(k))$. Consequently, (5.3) guides the input variable x toward a minima for $F(x)$.

Ex 1: As an example, consider the following nonlinear unconstrained optimization problem:

$$F(x_1, x_2) = \frac{25}{30}x_1^4 - \frac{1}{6}x_1^3 - \frac{7}{3}x_1^2 + \frac{2}{3}x_1 + \frac{25}{30}x_2^4 - \frac{1}{6}x_2^3 - \frac{7}{3}x_2^2 + \frac{2}{3}x_2 \tag{5.6}$$

The three-dimensional plot of this function is depicted in Fig. 5.2(a). The contour plot of this function is also illustrated in Fig. 5.2(b). As can be seen from these figures, this problem has three local minimums and a global minimum, which are marked in the figures. The gradient of this function is derived as follows:

$$\nabla F(x_1, x_2) = \begin{bmatrix} \frac{10}{3}x_1^3 - \frac{1}{2}x_1^2 - \frac{14}{3}x_1 + \frac{2}{3} \\ \frac{10}{3}x_2^3 - \frac{1}{2}x_2^2 - \frac{14}{3}x_2 + \frac{2}{3} \end{bmatrix} \tag{5.7}$$

In order to optimize this function, the following iterative equation can be used:

$$\begin{bmatrix} x_1(k+1) \\ x_2(k+1) \end{bmatrix} = \begin{bmatrix} x_1(k) \\ x_2(k) \end{bmatrix} - \alpha \begin{bmatrix} \frac{10}{3}x_1^3(k) - \frac{1}{2}x_1^2(k) - \frac{14}{3}x_1(k) + \frac{2}{3}(k) \\ \frac{10}{3}x_2^3(k) - \frac{1}{2}x_2^2(k) - \frac{14}{3}x_2(k) + \frac{2}{3}(k) \end{bmatrix} \tag{5.8}$$

Figure 5.3 shows the evolution of the input vector when the initial conditions are selected as $x(0) = [0,\ -0.2]^{\mathrm{T}}$ and the learning rate α is selected as 0.1. As can be seen from this figure, the algorithm converges to the global minimum of the function. This is a successful implementation of the GD on a complex nonlinear multi-variable function. However, in order to show the sensitivity of GD to initial conditions, another set of experiments are carried out. In the new experiments, three different initial conditions are selected as being equal to $[0.5,\ 0.2]^{\mathrm{T}}$, $[2,\ -2]^{\mathrm{T}}$, and $[-0.5,\ 0.5]^{\mathrm{T}}$ and the learning rate α is selected as being equal to 0.1. The results of the simulations are depicted in Fig. 5.4. As can be seen

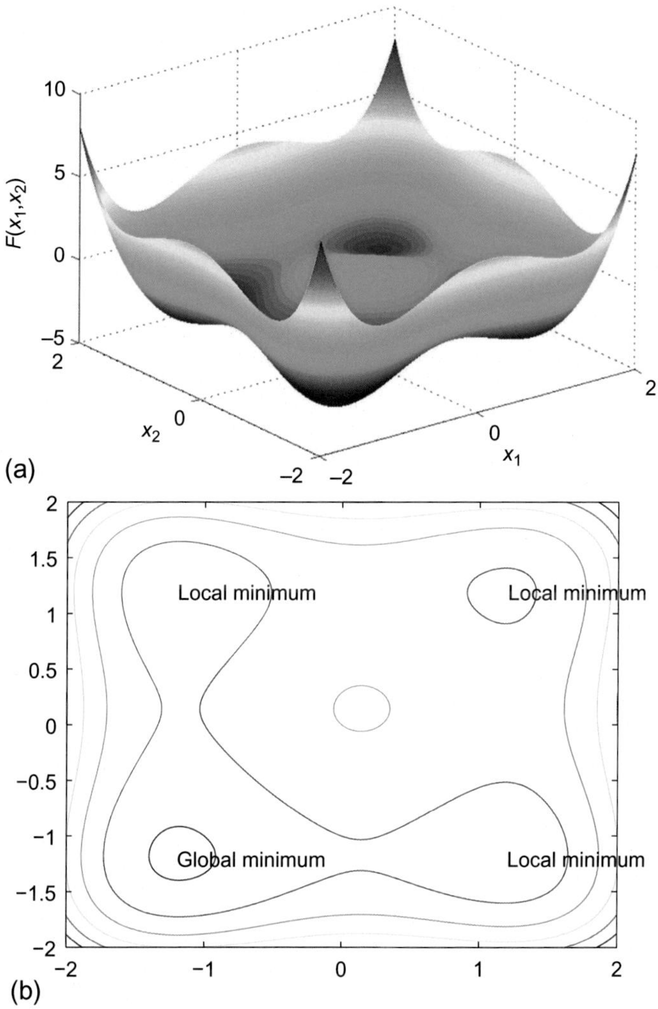

Figure 5.2 (a) Three-dimensional plot of the function in Ex. 1, (b) contour plot of the function given in Ex. 1.

from Fig. 5.4, the GD algorithm gets stuck in local minimums for all the aforementioned initial conditions. Thus, the conclusion of these simulations is that GD algorithm is quite sensitive to the initial conditions.

Ex 2: In order to show the effect of the selection of the learning rate, another experiment is conducted. Consider the following one-dimensional unconstraint function to be optimized:

$$F(x) = x^2 \tag{5.9}$$

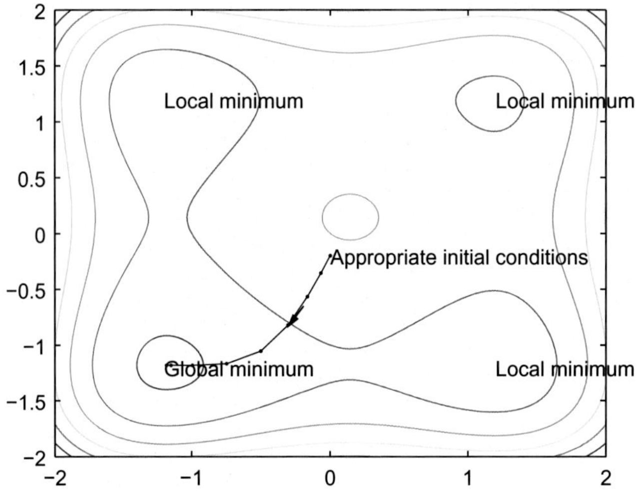

Figure 5.3 Result of applying GD to $F(x)$ when appropriate initial conditions are selected in Ex 1.

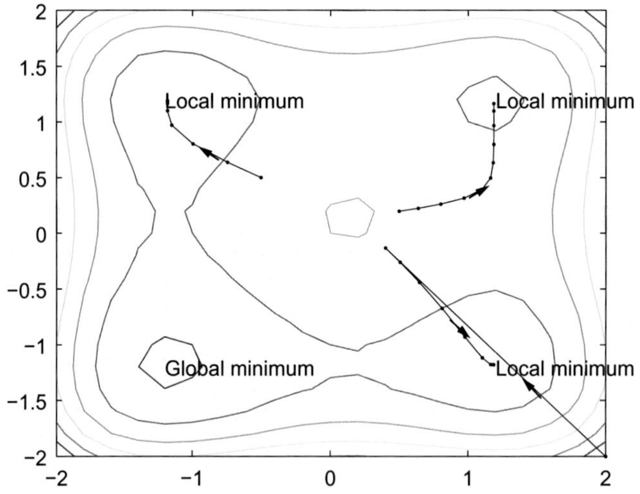

Figure 5.4 Result of applying GD to $F(x)$ when inappropriate initial conditions are selected in Ex 1.

The gradient of this function is achieved as follows:

$$\nabla F(x) = 2x \tag{5.10}$$

Hence, the GD algorithm update rule can be obtained as follows:

$$x(k+1) = x(k) - 2\alpha x(k) \tag{5.11}$$

In order to investigate the effect of the learning rate selection on the optimization process, two experiments are carried out with two different learning rates. In the first case, the initial conditions are selected as being equal to 2.5 and the learning rate is selected as $\alpha = 0.3$. It is observed that the minimum of the function can be found using the GD algorithm. On the other hand, in the second experiment, the same initial condition is used with the learning rate of 1.1. Figure 5.5 shows the evolution of the solutions when the learning rate is equal to 1.1. As can be seen from the figure, even if the GD algorithm gives the correct moving direction to optimize the function, the algorithm diverges and fails to find the optimum of the function since the step-size is chosen too large. In this chapter, we do not provide another experiment showing the disadvantage of choosing

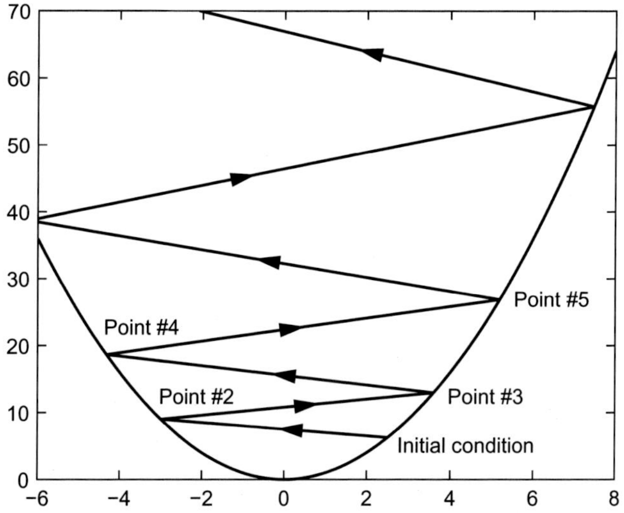

Figure 5.5 Result of applying GD to $F(x) = x^2$ when inappropriate value is selected for the learning rate in Ex. 2.

a very small learning rate. However, it is a well-known fact that such a very small selection of the learning rate increases the probability of entrapment in a local minimum. The overall finding is that neither a too small nor a too large learning rate is appropriate for the GD algorithm.

5.2.2 Newton and Gauss-Newton Optimization Algorithms

Different from the GD algorithm where we use only the first derivatives, Newton's optimization method is a second-order optimization method that also uses the second derivative, Hessian, of the function. Since the Hessian is taken into the account in this algorithm, it is expected that it gives better results when compared to an algorithm that only uses first derivatives. Moreover, since no learning rate exists in its original version, it encounters fewer instability issues. The second-order Taylor expansion of $F(x)$ around $x(k)$ is given as:

$$\begin{aligned} F(x) \approx G(x) = F(x(k)) + (\nabla F(x(k)))^{\mathrm{T}}(x - x(k)) \\ + \frac{1}{2}(x - x(k))^{\mathrm{T}} H(x(k))(x - x(k)) \end{aligned} \tag{5.12}$$

where $H(x)$ is the Hessian matrix of $F(x)$. Not only does (5.12) give the first order approximate of $F(x)$ but its second-order term is also taken into account. Thus, the approximation is more exact when it is compared to the first-order approximation. The second-order Taylor expansion of $F(x)$, which is represented by (5.12), is a quadratic function and it is possible to find its minima by solving $\nabla G(x) = 0$. The gradient of $G(x)$ is obtained as follows:

$$\nabla G(x) = \nabla F(x(k)) + H(x(k))(x - x(k)) \tag{5.13}$$

The optimum step size for minimizing $G(x)$ is obtained as the solution of the following equation:

$$\nabla G(x) = \nabla F(x(k)) + H(x(k))(x - x(k)) = 0 \tag{5.14}$$

which results in:

$$x(k+1) = -H^{-1}(x(k))\nabla F(x(k)) \tag{5.15}$$

As can be seen from (5.15), there exists no learning rate in the formula resulting in this algorithm having a suggestion for this parameter. In other words, if we compare (5.15) with (5.3), it can be observed that the Hessian matrix acts as a learning rate, which is a matrix rather than a scalar value. However, the calculation of the Hessian matrix is a tedious work, especially when it comes to the estimation of the parameters of the premise part of a FNN. Moreover, the calculation of the inverse of the Hessian matrix is needed, which may have a high dimension. The complexity of matrix inversion may be as high as $O(n^3)$, in which n is the number of the rows of the matrix to be inverted.

Since higher-order terms are neglected in the Newton optimization algorithm, it may be modified as follows to improve its robustness:

$$x(k+1) = -\gamma(H(x(k)))^{-1}\nabla F(x(k)),\ 0 < \gamma < 1 \tag{5.16}$$

Since the calculation of the Hessian matrix is complex as mentioned earlier, this matrix is approximated in order to reduce the complexity of the Newton optimization method.

Consider the cost function $F(x)$ to be in the following form:

$$F(x) = e^{\mathrm{T}}(x)e(x) \tag{5.17}$$

where $e(x) \in \mathbb{R}^{m\times 1}$ is a vectorial function of x that represents the difference between the desired value of the FNN and its output. The gradient of $F(x)$ is derived as follows:

$$\nabla F(x) = J^{\mathrm{T}}(x)e(x) \tag{5.18}$$

in which J is the Jacobian matrix and is defined as follows:

$$J(x) = \begin{bmatrix} \frac{\partial e_1(x)}{\partial x_1} & \frac{\partial e_1(x)}{\partial x_2} & \cdots & \frac{\partial e_1(x)}{\partial x_n} \\ \frac{\partial e_2(x)}{\partial x_1} & \frac{\partial e_2(x)}{\partial x_2} & \cdots & \frac{\partial e_2(x)}{\partial x_n} \\ \vdots & \vdots & \ddots & \vdots \\ \frac{\partial e_m(x)}{\partial x_1} & \frac{\partial e_m(x)}{\partial x_2} & \cdots & \frac{\partial e_m(x)}{\partial x_n} \end{bmatrix}$$

Furthermore, the Hessian matrix is obtained as follows:

$$\nabla^2 F(x) = J^{\mathrm{T}}(x)J(x) + S(x) \tag{5.19}$$

where:

$$S(x) = \sum_{i=1}^{m} e_i(x)\nabla^2 e_i(x)$$

and $e_i(x)$ is the ith element of the vector $e(x)$. In this way, the terms that include the first-order gradient are separated from the terms that include the second-order gradient. In other words, $J^{\mathrm{T}}(x)J(x)$ includes the first-order gradient while $S(x)$ capsulates the second-order gradient terms. Since it is difficult to calculate the second-order gradient term $S(x)$, this term may be completely neglected. Hence, the update rule of (5.16) can be approximated as:

$$x(k+1) = x(k) - [J^{\mathrm{T}}(x(k))J(x(k))]^{-1}J^{\mathrm{T}}(x(k))e(x(k)) \tag{5.20}$$

This iterative method to find the optimal value of $F(x)$ is called the Gauss-Newton optimization method. It is further possible to derive the formulation of the Gauss-Newton optimization method using a different approach. Consider the same cost function as for the previous approach as follows:

$$F(x) = e^{\mathrm{T}}(x)e(x) \tag{5.21}$$

It is possible to use the first-order Taylor expansion for $e(x)$ around $x(k)$ as follows:

$$e(x) \approx e(x(k)) + (J(x(k)))^{\mathrm{T}}(x - x(k)) \tag{5.22}$$

where $J(x)$ is Jacobian matrix. The term $x(k+1)$ is the solution to the cost function defined by (5.21). Hence, we have the following equation:

$$\begin{aligned}
x(x+1) &= arg\ \min_{x\in\mathbb{R}^n} e^{\mathrm{T}}(x)e(x) \\
&\approx \min_{x\in\mathbb{R}^n} \left[e(x(k)) + (J(x(k)))^{\mathrm{T}}(x - x(k))\right]^{\mathrm{T}} \\
&\quad \times \left[e(x(k)) + (J(x(k)))^{\mathrm{T}}(x - x(k))\right] \\
&= \min_{x\in\mathbb{R}^n} \{e^{\mathrm{T}}(x(k))e(x(k)) \\
&\quad + 2(x - x(k))^{\mathrm{T}}J(x(k))e(x(k)) \\
&\quad + (x - x(k))^{\mathrm{T}}J(x(k))(J(x(k)))^{\mathrm{T}}(x - x(k))
\end{aligned} \tag{5.23}$$

The above equation is a quadratic equation and its optimal point can be easily found by making its gradient equal to zero. Hence, its minimum point is obtained by the following equation:

$$x(k+1) = x(k) - [J(x(k))(J(x(k)))^{\mathrm{T}}]^{-1}J(x(k))e(x(k)) \tag{5.24}$$

5.2.3 LM Algorithm

As can be seen from (5.24), it is possible that $J(x(k))J^{\mathrm{T}}(x(k))$ may become singular or close to singular. In order to solve this problem, the original update rule for the Gauss-Newton algorithm is modified as follows:

$$x(k+1) = x(k) - [J(x(k))J^{\mathrm{T}}(x(k)) + \mu(k)I]^{-1}J(x(k))e(x(k)) \tag{5.25}$$

where I is a identity matrix having an appropriate size and $\mu(k)$ has a positive value, which avoids the inversion of a singular matrix. As can be seen from (5.25), if $\mu(k)$ is chosen as equal to zero, this equation is quite similar to the one in the Gauss-Newton case. Moreover, if a large value for $\mu(k)$ is chosen, $J(x(k))J^{\mathrm{T}}(x(k))$ can be ignored and (5.25) changes to a simple GD algorithm. Therefore, it is suggested to begin with a small $\mu(k)$ to use Gauss-Newton and speed-up convergence. In this way, the algorithm will be less sensitive to initial values considered for $x(k)$. After some iterations, it is suggested to increase $\mu(k)$ to avoid singularity. The pseudo code for the LM algorithm, which includes an appropriate selection for μ, can be summarized as in Fig. 5.6 [2].

5.2.4 Gradient Descent Algorithm with an Adaptive Learning Rate

It is possible to consider having an adaptive learning rate in the GD algorithm. Possible guidelines for increasing or decreasing the learning rate are as follows:

1. If a sign change in the solution does not occur in multiple consecutive iterations, the learning rate is possibly chosen too small, resulting in the algorithm converging too slow. In this case, it is wise to increase the value of the learning rate to increase the convergence speed.
2. If the sign of $\partial F/\partial x_j$ changes in multiple consecutive iterations, the algorithm is possibility oscillating around a local solution. In this case, it is wise to decrease the learning rate.

The adaptive learning rate is mostly used in batch training algorithms [3].

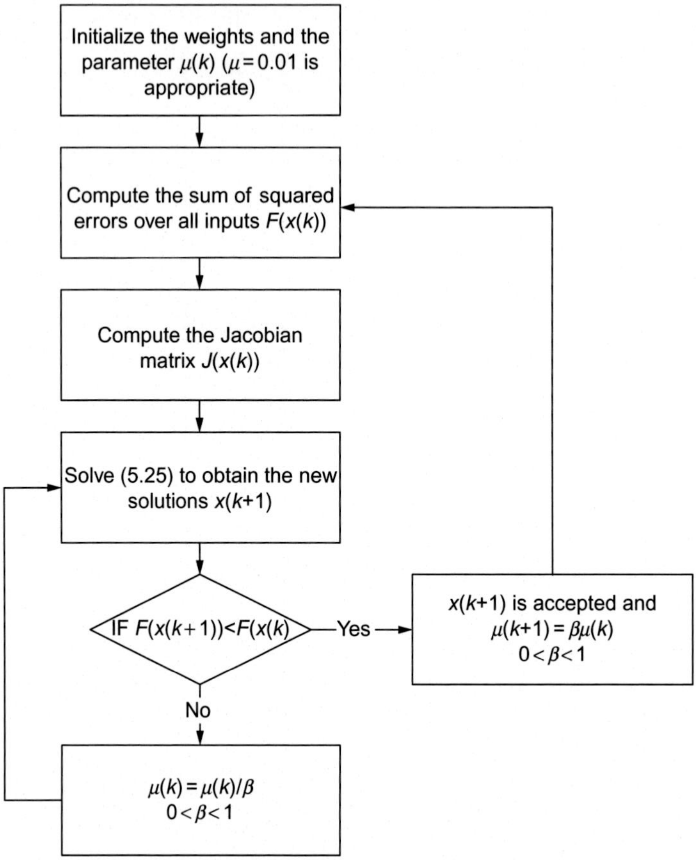

Figure 5.6 Flowchart of the LM algorithm.

5.2.5 GD Algorithm with a Momentum Term

Another possible solution to decreasing the risk of instability is using a momentum term. The GD algorithm with a momentum term is defined as follows:

$$\Delta x(k) = x(k+1) - x(k) = -\alpha \frac{\partial F(x(k))}{\partial x(k)} + \beta \Delta x(k-1) \tag{5.26}$$

In order to see how the momentum term is affecting the training algorithm, an analysis is done here. If we consider that x is a one-dimensional variable, we can take the Z-transform of $\Delta x(k)$, so that the following equation is derived:

$$\Delta X(z) = -\alpha \frac{\partial F(x(k))}{\partial x(k)} + z^{-1} \beta \Delta X(z) \tag{5.27}$$

Therefore:

$$\Delta X(z) = \frac{-\alpha \frac{\partial F(x(k))}{\partial x(k)}}{1 - \beta z^{-1}} \tag{5.28}$$

in which $\frac{1}{1-\beta z^{-1}}$ acts as a low-pass filter that filters $-\alpha \frac{\partial F(x(k))}{\partial x(k)}$ and reduces oscillations in $x(k)$ considerably. Furthermore, in order to have a stable filter, the value of β must be selected as $0 < \beta < 1$.

5.3 GRADIENT DESCENT BASED LEARNING ALGORITHMS FOR TYPE-2 FUZZY NEURAL NETWORKS

5.3.1 Consequent Part Parameters

As mentioned earlier, the output of an interval T2FNN is computed as follows:

$$y_N = q \sum_{r=1}^{N} f_r \underline{\tilde{w}} + (1 - q) \sum_{r=1}^{N} f_r \overline{\tilde{w}} \tag{5.29}$$

where:

$$f_r = a_{ri} x_i + b_r, \quad \forall r = 1, 2, \ldots, N, \quad \text{and} \quad i = 1, 2, \ldots, n$$

As can be seen from the equation above, the output of a FNN, either type-1 or type-2, is linear with respect to the parameters of the consequent part of the FNN. Hence, it is quite easy to derive the GD-based adaptation laws for the parameters of the consequent part. The cost function is rewritten for ease of reference:

$$E = \frac{1}{2} e^2 = \frac{1}{2} (y - y_N)^2 \tag{5.30}$$

The GD-based parameter tuning rules for the consequent part are derived as follows:

$$a_{ri}(k+1) = a_{ri}(k) + \alpha \left(q \underline{\tilde{w}} + (1 - q) \overline{\tilde{w}} \right) x_i(k) e(k), \ 0 < \alpha \tag{5.31}$$

$$b_r(k+1) = b_r(k) + \alpha\left(q\underline{\tilde{w}} + (1-q)\overline{\tilde{w}}\right)e(k), \quad 0 < \alpha \tag{5.32}$$

$$q(k+1) = q(k) + \alpha\left(\sum_{r=1}^{N} f_r(\underline{\tilde{w}} - \overline{\tilde{w}})\right)e(k), \quad 0 < \alpha \tag{5.33}$$

where α is the learning rate for the consequent part parameters. The parameter q has a bounded value by definition, and it should be kept in the interval $[0, \ 1]$. As can be seen from these equations, GD-based adaptation rules for the parameters of the consequent part are straightforward and have closed form. Note that the learning rates of the parameters of the consequent part do not need to be equal and thus may differ from one parameter to another. It should also be noted that in some studies, the GD-based training method is used to tune only the parameters of the consequent part, and the parameters of the the antecedent part are kept constant. Since the sensitivity of the output of a FNN to the parameters of the consequent part are more than that of antecedent part, such approaches may result in fairly accurate performance, but with fast training benefiting from simple adaptation laws. However, in order to design a more precise FNN, the parameters of the antecedent part should also be tuned. Unfortunately, this is a more complex task when compared to the tuning of the consequent part.

5.3.2 Premise Part Parameters

In the previous section, it was already stated that the calculation of the partial derivative of the T2FNN output with respect to the parameters of the antecedent part is more difficult than that of the consequent part. Since there exists a normalization layer in the structure of a T2FNN, the parameters of the premise part of each MF, no matter if it has participated in a rule or not, exists at least in the denominator of the output of the normalization layer. Hence, there are no closed forms for computing the partial derivative of the T2FNN output with respect to its premise part. In this chapter, the adaptation laws for a T2FNN, which has only two inputs, is considered.

Consider a T2FNN with two inputs and I number of MFs for the first input and J number of MFs for the second input. The upper MFs for the first input x_1 are considered to be $\overline{\mu}_{11}(x_1), \overline{\mu}_{12}(x_1), \ldots,$ and $\overline{\mu}_{1I}(x_1)$, and its corresponding lower MFs are considered to be $\underline{\mu}_{11}(x_1), \underline{\mu}_{12}(x_1), \ldots$ and $\underline{\mu}_{1I}(x_1)$. The upper MFs for the second input (x_2) are considered to be $\overline{\mu}_{21}(x_2), \overline{\mu}_{22}(x_2), \ldots,$ and $\overline{\mu}_{2J}(x_2)$, and its corresponding lower MFs are considered to be $\underline{\mu}_{21}(x_2), \underline{\mu}_{22}(x_2), \ldots,$ and $\underline{\mu}_{2J}(x_2)$. The terms $\underline{\mu}_{11}(x_2)$, $\overline{\mu}_{11}(x_1)$ are considered to be defined as follows:

$$\underline{\mu}_{11}(x_1) = exp\left(-\left(\frac{x_1 - c_{11}}{\underline{\sigma}_{11}}\right)^2\right) \tag{5.34}$$

$$\overline{\mu}_{11}(x_1) = exp\left(-\left(\frac{x_1 - c_{11}}{\overline{\sigma}_{11}}\right)^2\right) \tag{5.35}$$

The upper and the lower rules of the system are derived as follows:

$$\begin{aligned}
\underline{w}_1 &= \underline{\mu}_{11}(x_1)\underline{\mu}_{21}(x_2), \ldots, \underline{w}_{1\times J} = \underline{\mu}_{11}(x_1)\underline{\mu}_{2J}(x_2) \\
\underline{w}_{1\times J+1} &= \underline{\mu}_{12}(x_1)\underline{\mu}_{21}(x_2), \ldots, \underline{w}_{2\times J} = \underline{\mu}_{12}(x_1)\underline{\mu}_{2J}(x_2) \\
&\vdots \\
\underline{w}_{(I-1)\times J+1} &= \underline{\mu}_{1I}(x_1)\underline{\mu}_{21}(x_2), \ldots, \underline{w}_{I\times J} = \underline{\mu}_{1I}(x_1)\underline{\mu}_{2J}(x_2)
\end{aligned} \tag{5.36}$$

and

$$\begin{aligned}
\overline{w}_1 &= \overline{\mu}_{11}(x_1)\overline{\mu}_{21}(x_2), \ldots, \overline{w}_{1\times J} = \overline{\mu}_{11}(x_1)\overline{\mu}_{2J}(x_2) \\
\overline{w}_{1\times J+1} &= \overline{\mu}_{12}(x_1)\overline{\mu}_{21}(x_2), \ldots, \overline{w}_{2\times J} = \overline{\mu}_{12}(x_1)\overline{\mu}_{2J}(x_2) \\
&\vdots \\
\overline{w}_{(I-1)\times J+1} &= \overline{\mu}_{1I}(x_1)\overline{\mu}_{21}(x_2), \ldots, \overline{w}_{I\times J} = \overline{\mu}_{1I}(x_1)\overline{\mu}_{2J}(x_2)
\end{aligned} \tag{5.37}$$

The output of the T2FNN is obtained as follows:

$$y_N = q \sum_{r=1}^{N=I\times J} f_r\tilde{\underline{w}}_r + (1-q) \sum_{r=1}^{N=I\times J} f_r\tilde{\overline{w}}_r \tag{5.38}$$

where:

$$\tilde{\underline{w}}_r = \frac{\underline{w}_r}{\sum_{l=1}^{N} \underline{w}_l}, \quad \tilde{\overline{w}}_r = \frac{\overline{w}_r}{\sum_{l=1}^{N} \overline{w}_l} \tag{5.39}$$

The chain rule to obtain the partial derivative of $F = \frac{1}{2}e^2$ with respect to a sample parameter c_1 is given in (5.40). F is the cost function to be optimized, which is itself a function of $e = y - y_N$, in which y_N is the output of T2FNN and y is its corresponding target value. This partial derivative is obtained as follows:

$$\frac{\partial F}{\partial c_{11}} = \frac{\partial F}{\partial e} . \frac{\partial e}{\partial y_N} . \frac{\partial y_N}{\partial c_{11}} \tag{5.40}$$

and further:

$$\frac{\partial y_N}{\partial c_{11}} = \sum_{r=1}^{N} \sum_{j=1}^{J} \left(\frac{\partial y_N}{\partial \underline{\tilde{w}}_r} \frac{\partial \underline{\tilde{w}}_r}{\partial \underline{w}_j} \frac{\partial \underline{w}_j}{\partial \underline{\mu}_{11}} \frac{\partial \underline{\mu}_{11}}{\partial c_{11}} \right) + \sum_{r=1}^{N} \sum_{j=1}^{J} \left(\frac{\partial y_N}{\partial \tilde{\overline{w}}_r} \frac{\partial \tilde{\overline{w}}_r}{\partial \overline{w}_j} \frac{\partial \overline{w}_j}{\partial \overline{\mu}_{11}} \frac{\partial \overline{\mu}_{11}}{\partial c_{11}} \right) \tag{5.41}$$

where:

$$\begin{aligned} \frac{\partial y_N}{\partial \underline{\tilde{w}}_r} &= f_r, r = 1, \ldots, N \\ \frac{\partial \underline{w}_i}{\partial \underline{\mu}_{11}} &= \underline{\mu}_{2j}, j = 1, \ldots, J \\ \frac{\partial \underline{\mu}_{11}}{\partial c_{11}} &= -(exp(-(x_1 - c_{11})^2/(2\underline{\sigma}_{11}^2))(2c_{11} - 2x_1))/(2\sigma_{11}^2) \end{aligned} \tag{5.42}$$

and similarly, we have the following equations:

$$\begin{aligned} \frac{\partial y_N}{\partial \tilde{\overline{w}}_r} &= f_r, r = 1, \ldots, N \\ \frac{\partial \overline{w}_i}{\partial \overline{\mu}_{11}} &= \overline{\mu}_{2j}, j = 1, \ldots, J \\ \frac{\partial \overline{\mu}_{11}}{\partial c_{11}} &= -(exp(-(x_1 - c_{11})^2/(2\overline{\sigma}_{11}^2))(2c_{11} - 2x_1))/(2\sigma_{11}^2) \end{aligned} \tag{5.43}$$

However, the partial derivatives of $\underline{\tilde{w}}_r$, $r = 1, \ldots, N$ w.r.t. $\underline{w}_j$, $j = 1, \ldots, J$ and the partial derivatives of $\tilde{\overline{w}}_r$, $r = 1, \ldots, N$ w.r.t. $\overline{w}_j$, $j = 1, \ldots, J$ are multi-rule. If $\overline{w}_j$ appears only in the denominator of $\tilde{\overline{w}}_r$, the partial derivatives of $\tilde{\overline{w}}_r$'s w.r.t. $\overline{w}_j$'s are derived as follows:

$$\frac{\partial \tilde{\overline{w}}_r}{\partial \overline{w}_j} = -\frac{\tilde{\overline{w}}_r}{\sum_{l=1}^{N} \overline{w}_l} \tag{5.44}$$

However, in some cases $\overline{w}_j$ appears both in the numerator and denominator of $\tilde{\overline{w}}_r$. In these cases, the partial derivatives of $\tilde{\overline{w}}_r$'s w.r.t. $\overline{w}_j$'s are calculated as follows:

$$\frac{\partial \tilde{\overline{w}}_r}{\partial \overline{w}_j} = \frac{\partial \tilde{\overline{w}}_r}{\partial \overline{w}_r} = \frac{1 - \tilde{\overline{w}}_r}{\sum_{i=l}^{N} \overline{w}_l} \tag{5.45}$$

Similar calculations apply for the partial derivatives of $\tilde{\underline{w}}_r$ w.r.t. $\underline{w}_j$. This ends the calculation of the partial derivative of the T2FNN output w.r.t. c_{11}, which is one of the parameters of the antecedent part of the T2FNN. As can be seen from these calculations, the partial derivative of all the normalized rules w.r.t. c_{11} exists in the calculation of the partial derivative of y_N w.r.t. c_{11}, which is generally non-zero. This is the main reason the computation of the partial derivative with respect to the parameters of the consequent part is hard to derive. Interested readers are encouraged to calculate the rest of the partial derivatives of y_N w.r.t. the parameters of the antecedent part and complete the GD learning rules.

5.3.3 Variants of the Back-Propagation Algorithm for Training the T2FNNs

Having completed the calculation of the partial derivative of y_N w.r.t. the parameters of T2FNN, any algorithm such as basic GD, momentum term GD and LM algorithms may be applied to the parameter update rules of a T2FNN. However, because of the high complexity of the Newton's optimization algorithm, it has never been applied to the training of T2FNN.

5.4 STABILITY ANALYSIS

GD-based training methods suffer from lack of stability analysis. In this section, a stability analysis of the GD algorithm is considered. However, note that the Taylor expansion with only the first-order term is considered, which makes the analysis local and valid for only small changes in the solution. Similar stability analysis for the GD training of FNNs can be found in Refs. [4, 5].

5.4.1 Stability Analysis of GD for Training of T2FNN

In order to analyze the stability of the GD-based training algorithm, the following discrete time Lyapunov function is considered:

$$V(k) = e^2(k) \tag{5.46}$$

By using the Lyapunov theory, in order to prove the stability of a discrete time system, it is required that its time difference is negative. Thus, in the following equation, the time difference of the Lyapunov function is considered:

$$\Delta V(k) = V(k+1) - V(k) = e^2(k+1) - e^2(k) \tag{5.47}$$

By using the first-order Taylor expansion of $e(k)$, we have the following equation:

$$\begin{aligned}
\Delta V(k) &= \Delta e(k)\left(e(k) + \frac{1}{2}\Delta e(k)\right) \\
&= \left[\left(\frac{\partial e(k)}{\partial \overline{\sigma}}\right)^{\mathrm{T}} \Delta\overline{\sigma} + \left(\frac{\partial e(k)}{\partial \underline{\sigma}}\right)^{\mathrm{T}} \Delta\underline{\sigma} + \left(\frac{\partial e(k)}{\partial c}\right)^{\mathrm{T}} \Delta c \right.\\
&\quad \left. + \left(\frac{\partial e(k)}{\partial f}\right)^{\mathrm{T}} \Delta f\right] \times \left[e(k) + \frac{1}{2}\left(\frac{\partial e(k)}{\partial \overline{\sigma}}\right)^{\mathrm{T}} \Delta\overline{\sigma}\right.\\
&\quad \left. + \frac{1}{2}\left(\frac{\partial e(k)}{\partial \underline{\sigma}}\right)^{\mathrm{T}} \Delta\underline{\sigma} + \frac{1}{2}\left(\frac{\partial e(k)}{\partial c}\right)^{\mathrm{T}} \Delta c + \frac{1}{2}\left(\frac{\partial e(k)}{\partial f}\right)^{\mathrm{T}} \Delta f\right]
\end{aligned} \tag{5.48}$$

Therefore:

$$\begin{aligned}
\Delta V(k) = &-\frac{1}{2}e^2(k)\left[\left(\frac{\partial y_{NN}(k)}{\partial \overline{\sigma}}\right)^{\mathrm{T}} \eta_{\overline{\sigma}} \frac{\partial y_{NN}(k)}{\partial \overline{\sigma}}\right.\\
&+ \left(\frac{\partial y_{NN}(k)}{\partial \underline{\sigma}}\right)^{\mathrm{T}} \eta_{\underline{\sigma}} \frac{\partial y_{NN}(k)}{\partial \underline{\sigma}} + \left(\frac{\partial y_{NN}(k)}{\partial c}\right)^{\mathrm{T}} \eta_{c} \frac{\partial y_{NN}(k)}{\partial c}\\
&\left.+ \left(\frac{\partial y_{NN}(k)}{\partial f}\right)^{\mathrm{T}} \eta_{f} \frac{\partial y_{NN}(k)}{\partial f}\right] \times \left[2 - \left(\frac{\partial y_{NN}(k)}{\partial \overline{\sigma}}\right)^{\mathrm{T}} \eta_{\overline{\sigma}} \frac{\partial y_{NN}(k)}{\partial \overline{\sigma}}\right.\\
&- \left(\frac{\partial y_{NN}(k)}{\partial \underline{\sigma}}\right)^{\mathrm{T}} \eta_{\underline{\sigma}} \frac{\partial y_{NN}(k)}{\partial \underline{\sigma}} - \left(\frac{\partial y_{NN}(k)}{\partial c}\right)^{\mathrm{T}} \eta_{c} \frac{\partial y_{NN}(k)}{\partial c}\\
&\left.- \left(\frac{\partial y_{NN}(k)}{\partial f}\right)^{\mathrm{T}} \eta_{f} \frac{\partial y_{NN}(k)}{\partial f}\right]
\end{aligned} \tag{5.49}$$

and further:

$$\Delta V(k) = -\frac{1}{2}e^2(k)\left[\eta_{\overline{\sigma}}\left\|\frac{\partial y_{NN}(k)}{\partial \overline{\sigma}}\right\|^2 + \eta_{\underline{\sigma}}\left\|\frac{\partial y_{NN}(k)}{\partial \underline{\sigma}}\right\|^2 + \eta_c\left\|\frac{\partial y_{NN}(k)}{\partial c}\right\|^2 + \eta_f\left\|\frac{\partial y_{NN}(k)}{\partial f}\right\|^2\right] \times \left[2 - \eta_{\overline{\sigma}}\left\|\frac{\partial y_{NN}(k)}{\partial \overline{\sigma}}\right\|^2 - \eta_{\underline{\sigma}}\left\|\frac{\partial y_{NN}(k)}{\partial \underline{\sigma}}\right\|^2 - \eta_c\left\|\frac{\partial y_{NN}(k)}{\partial c}\right\|^2 - \eta_f\left\|\frac{\partial y_{NN}(k)}{\partial f}\right\|^2\right] \quad (5.50)$$

which implies that in order to have $\Delta V(k) < 0$, we must have the following inequalities:

$$0 < \eta_{\overline{\sigma}} < \frac{1}{2\left\|\frac{\partial y_{NN}}{\partial \overline{\sigma}}\right\|^2}, 0 < \eta_{\underline{\sigma}} < \frac{1}{2\left\|\frac{\partial y_{NN}}{\partial \underline{\sigma}}\right\|^2}$$
$$0 < \eta_c < \frac{1}{2\left\|\frac{\partial y_{NN}}{\partial c}\right\|^2}, 0 < \eta_f < \frac{1}{2\left\|\frac{\partial y_{NN}}{\partial f}\right\|^2} \quad (5.51)$$

The above equations give an adaptive selection for the adaptive learning rates that stabilizes the GD-based learning algorithm.

5.4.2 Stability Analysis of the LM for Training of T2FNN

The stability analysis of LM method for the training of FNN has been previously considered in Ref. [6]. However, the proof of the stability in Ref. [6] requires the eigenvalues of $J(x(t))J^{\mathrm{T}}(x(t))$ to be computed, which is a very tedious work. The complexity of the eigenvalues computation of the matrices is at least $2n^3/3 + O(n^2)$. Alternatively, the benefit of the stability analysis in this chapter is that it does not require any eigenvalue to be computed, and hence it is much simpler. As a beginning point, the following Lyapunov function is considered:

$$V(k) = e^2(k) \quad (5.52)$$

In order to have a stable training algorithm, $\Delta V(k)$ must be less than zero. The following equation is achieved for $\Delta V(k)$:

$$\Delta V(k) = V(k+1) - V(k) = e^2(k+1) - e^2(k) \tag{5.53}$$

By using the first-order Taylor expansion of $e(k)$, we have the following equation:

$$\begin{aligned}\Delta V(k) &= \Delta e(k)\left(e(k) + \frac{1}{2}\Delta e(k)\right)\\ &\approx \left[\left(\frac{\partial e(k)}{\partial \overline{\sigma}}\right)^{\mathrm{T}} \Delta \overline{\sigma} + \left(\frac{\partial e(k)}{\partial \underline{\sigma}}\right)^{\mathrm{T}} \Delta \underline{\sigma} + \left(\frac{\partial e(k)}{\partial c}\right)^{\mathrm{T}} \Delta c\right.\\ &\left.+\left(\frac{\partial e(k)}{\partial f}\right)^{\mathrm{T}} \Delta f\right] \times \left[e(k) + \frac{1}{2}\left(\frac{\partial e(k)}{\partial \overline{\sigma}}\right)^{\mathrm{T}} \Delta \overline{\sigma}\right.\\ &\left.+\frac{1}{2}\left(\frac{\partial e(k)}{\partial \underline{\sigma}}\right)^{\mathrm{T}} \Delta \underline{\sigma} + \frac{1}{2}\left(\frac{\partial e(k)}{\partial c}\right)^{\mathrm{T}} \Delta c + \frac{1}{2}\left(\frac{\partial e(k)}{\partial f}\right)^{\mathrm{T}} \Delta f\right]\end{aligned} \tag{5.54}$$

so that:

$$\begin{aligned}\left(\frac{\partial e(k)}{\partial \overline{\sigma}}\right)^{\mathrm{T}} \Delta \overline{\sigma} &= \left(\frac{\partial e(k)}{\partial \overline{\sigma}}\right)^{\mathrm{T}} \left(\frac{\partial e(k)}{\partial \overline{\sigma}} \left(\frac{\partial e(k)}{\partial \overline{\sigma}}\right)^{\mathrm{T}} + \mu_{\overline{\sigma}} I\right)^{-1} \left(\frac{\partial e}{\partial \overline{\sigma}}\right) e(k)\\ &= \left(\frac{\partial y_{NN}(k)}{\partial \overline{\sigma}}\right)^{\mathrm{T}} \left(\frac{\partial y_{NN}(k)}{\partial \overline{\sigma}} \left(\frac{\partial y_{NN}(k)}{\partial \overline{\sigma}}\right)^{\mathrm{T}}\right.\\ &\left.+\mu_{\overline{\sigma}} I\right)^{-1} \left(\frac{\partial y_{NN}}{\partial \overline{\sigma}}\right) e(k)\end{aligned} \tag{5.55}$$

It is possible to define an auxiliary variable Ξ as:

$$\Xi = \left(\frac{\partial y_{NN}(k)}{\partial \overline{\sigma}} \left(\frac{\partial y_{NN}(k)}{\partial \overline{\sigma}}\right)^{\mathrm{T}} + \mu_{\overline{\sigma}} I\right)^{-1} \tag{5.56}$$

Lemma 5.1 (Matrix Inversion Lemma). *Let A, C, and $C^{-1} + DA^{-1}B$ be non-singular square matrices. Then $A + BCD$ is invertible, and [7]:*

$$(A + BCD)^{-1} = A^{-1} - A^{-1}B(C^{-1} + DA^{-1}B)^{-1}DA^{-1} \tag{5.57}$$

Using Lemma 5.1. it is possible to write Ξ as follows:

$$\begin{aligned}\Xi &= \mu_{\overline{\sigma}}^{-1} I - \mu_{\overline{\sigma}}^{-1} \frac{\partial y_{NN}}{\partial \overline{\sigma}} \left(I + \mu_{\overline{\sigma}} \left(\frac{\partial y_{NN}}{\partial \overline{\sigma}} \right)^{\mathrm{T}} \frac{\partial y_{NN}}{\partial \overline{\sigma}} \right)^{-1} \left(\frac{\partial y_{NN}}{\partial \overline{\sigma}} \right)^{\mathrm{T}} \mu_{\overline{\sigma}}^{-1} \\ &= \mu_{\overline{\sigma}}^{-1} I - \frac{\partial y_{NN}}{\partial \overline{\sigma}} \left(\mu_{\overline{\sigma}} I + \left(\frac{\partial y_{NN}}{\partial \overline{\sigma}} \right)^{\mathrm{T}} \frac{\partial y_{NN}}{\partial \overline{\sigma}} \right)^{-1} \left(\frac{\partial y_{NN}}{\partial \overline{\sigma}} \right)^{\mathrm{T}} \mu_{\overline{\sigma}}^{-1}\end{aligned} \tag{5.58}$$

and

$$\begin{aligned}\left(\frac{\partial e(k)}{\partial \overline{\sigma}} \right)^{\mathrm{T}} \Delta \overline{\sigma} &= \left(\frac{\partial y_{NN}}{\partial \overline{\sigma}} \right)^{\mathrm{T}} \Xi \frac{\partial y_{NN}}{\partial \overline{\sigma}} e(k) \\ &= \mu_{\overline{\sigma}}^{-1} \left\| \frac{\partial y_{NN}}{\partial \overline{\sigma}} \right\|^{2} e(k) - \mu_{\overline{\sigma}}^{-1} \left\| \frac{\partial y_{NN}}{\partial \overline{\sigma}} \right\|^{4} \\ &\quad \left(\mu_{\overline{\sigma}} + \left\| \frac{\partial y_{NN}}{\partial \overline{\sigma}} \right\|^{2} \right)^{-1} e(k)\end{aligned} \tag{5.59}$$

$\left(\frac{\partial e(k)}{\partial \underline{\sigma}} \right)^{\mathrm{T}} \Delta \underline{\sigma}$ is obtained as follows:

$$\begin{aligned}\left(\frac{\partial e(k)}{\partial \underline{\sigma}} \right)^{\mathrm{T}} \Delta \underline{\sigma} &= \mu_{\underline{\sigma}}^{-1} \left\| \frac{\partial y_{NN}}{\partial \underline{\sigma}} \right\|^{2} e(k) - \mu_{\underline{\sigma}}^{-1} \left\| \frac{\partial y_{NN}}{\partial \underline{\sigma}} \right\|^{4} \\ &\quad \left(\mu_{\underline{\sigma}} + \left\| \frac{\partial y_{NN}}{\partial \underline{\sigma}} \right\|^{2} \right)^{-1} e(k)\end{aligned} \tag{5.60}$$

Similarly we have:

$$\begin{aligned}\left(\frac{\partial e(k)}{\partial c} \right)^{\mathrm{T}} \Delta c &= \mu_{c}^{-1} \left\| \frac{\partial y_{NN}}{\partial c} \right\|^{2} e(k) - \mu_{c}^{-1} \left\| \frac{\partial y_{NN}}{\partial c} \right\|^{4} \\ &\quad \left(\mu_{c} + \left\| \frac{\partial y_{NN}}{\partial c} \right\|^{2} \right)^{-1} e(k)\end{aligned} \tag{5.61}$$

and:

$$\left(\frac{\partial e(k)}{\partial f}\right)^{\mathrm{T}} \Delta f = \mu_f^{-1} \left\|\frac{\partial y_{NN}}{\partial f}\right\|^2 e(k) - \mu_f^{-1} \left\|\frac{\partial y_{NN}}{\partial f}\right\|^4 \left(\mu_f + \left\|\frac{\partial y_{NN}}{\partial f}\right\|^2\right)^{-1} e(k) \tag{5.62}$$

Therefore, the following equation is obtained for the time difference of the Lyapunov function $V(k)$:

$$\begin{aligned}
\Delta V(k) = -\frac{1}{2}e^2(k) \Bigg(& \mu_{\overline{\sigma}}^{-1} \left\|\frac{\partial y_{NN}}{\partial \overline{\sigma}}\right\|^2 - \mu_{\overline{\sigma}}^{-1} \left\|\frac{\partial y_{NN}}{\partial \overline{\sigma}}\right\|^4 \left(\mu_{\overline{\sigma}} + \left\|\frac{\partial y_{NN}}{\partial \overline{\sigma}}\right\|^2\right)^{-1} \\
& + \mu_{\underline{\sigma}}^{-1} \left\|\frac{\partial y_{NN}}{\partial \underline{\sigma}}\right\|^2 - \mu_{\underline{\sigma}}^{-1} \left\|\frac{\partial y_{NN}}{\partial \underline{\sigma}}\right\|^4 \left(\mu_{\underline{\sigma}} + \left\|\frac{\partial y_{NN}}{\partial \underline{\sigma}}\right\|^2\right)^{-1} \\
& + \mu_c^{-1} \left\|\frac{\partial y_{NN}}{\partial c}\right\|^2 - \mu_c^{-1} \left\|\frac{\partial y_{NN}}{\partial c}\right\|^4 \left(\mu_c + \left\|\frac{\partial y_{NN}}{\partial c}\right\|^2\right)^{-1} \\
& + \mu_f^{-1} \left\|\frac{\partial y_{NN}}{\partial f}\right\|^2 - \mu_f^{-1} \left\|\frac{\partial y_{NN}}{\partial f}\right\|^4 \left(\mu_f + \left\|\frac{\partial y_{NN}}{\partial f}\right\|^2\right)^{-1} \Bigg) \\
\times \Bigg(2 & - \mu_{\overline{\sigma}}^{-1} \left\|\frac{\partial y_{NN}}{\partial \overline{\sigma}}\right\|^2 + \mu_{\overline{\sigma}}^{-1} \left\|\frac{\partial y_{NN}}{\partial \overline{\sigma}}\right\|^4 \left(\mu_{\overline{\sigma}} + \left\|\frac{\partial y_{NN}}{\partial \overline{\sigma}}\right\|^2\right)^{-1} \\
& - \mu_{\underline{\sigma}}^{-1} \left\|\frac{\partial y_{NN}}{\partial \underline{\sigma}}\right\|^2 + \mu_{\underline{\sigma}}^{-1} \left\|\frac{\partial y_{NN}}{\partial \underline{\sigma}}\right\|^4 \left(\mu_{\underline{\sigma}} + \left\|\frac{\partial y_{NN}}{\partial \underline{\sigma}}\right\|^2\right)^{-1} \\
& - \mu_c^{-1} \left\|\frac{\partial y_{NN}}{\partial c}\right\|^2 + \mu_c^{-1} \left\|\frac{\partial y_{NN}}{\partial c}\right\|^4 \left(\mu_c + \left\|\frac{\partial y_{NN}}{\partial c}\right\|^2\right)^{-1} \\
& - \mu_f^{-1} \left\|\frac{\partial y_{NN}}{\partial f}\right\|^2 + \mu_f^{-1} \left\|\frac{\partial y_{NN}}{\partial f}\right\|^4 \left(\mu_f + \left\|\frac{\partial y_{NN}}{\partial f}\right\|^2\right)^{-1} \Bigg)
\end{aligned} \tag{5.63}$$

Since:

$$0 \leq \left\| \frac{\partial y_{NN}}{\partial \overline{\sigma}} \right\|^2 \tag{5.64}$$

we have:

$$\mu_{\overline{\sigma}}^{-1} \left\| \frac{\partial y_{NN}}{\partial \overline{\sigma}} \right\|^4 \leq \left\| \frac{\partial y_{NN}}{\partial \overline{\sigma}} \right\|^2 + \mu_{\overline{\sigma}}^{-1} \left\| \frac{\partial y_{NN}}{\partial \overline{\sigma}} \right\|^4 \tag{5.65}$$

It is possible to multiply both sides by $\left(\mu_{\overline{\sigma}} + \| \frac{\partial y_{NN}}{\partial \overline{\sigma}} \|^2\right)^{-1}$ and we have:

$$\mu_{\overline{\sigma}}^{-1} \left\| \frac{\partial y_{NN}}{\partial \overline{\sigma}} \right\|^4 \left(\mu_{\overline{\sigma}} + \left\| \frac{\partial y_{NN}}{\partial \overline{\sigma}} \right\|^2\right)^{-1} \leq \mu_{\overline{\sigma}}^{-1} \left\| \frac{\partial y_{NN}}{\partial \overline{\sigma}} \right\|^2 \tag{5.66}$$

so that:

$$0 \leq \mu_{\overline{\sigma}}^{-1} \left\| \frac{\partial y_{NN}}{\partial \overline{\sigma}} \right\|^2 - \mu_{\overline{\sigma}}^{-1} \left\| \frac{\partial y_{NN}}{\partial \overline{\sigma}} \right\|^4 \left(\mu_{\overline{\sigma}} + \left\| \frac{\partial y_{NN}}{\partial \overline{\sigma}} \right\|^2\right)^{-1} \tag{5.67}$$

and using a similar analysis we get:

$$0 \leq \mu_{\underline{\sigma}}^{-1} \left\| \frac{\partial y_{NN}}{\partial \underline{\sigma}} \right\|^2 - \mu_{\underline{\sigma}}^{-1} \left\| \frac{\partial y_{NN}}{\partial \underline{\sigma}} \right\|^4 \left(\mu_{\underline{\sigma}} + \left\| \frac{\partial y_{NN}}{\partial \underline{\sigma}} \right\|^2\right)^{-1} \tag{5.68}$$

and:

$$0 \leq \mu_{c}^{-1} \left\| \frac{\partial y_{NN}}{\partial c} \right\|^2 - \mu_{c}^{-1} \left\| \frac{\partial y_{NN}}{\partial c} \right\|^4 \left(\mu_{c} + \left\| \frac{\partial y_{NN}}{\partial c} \right\|^2\right)^{-1} \tag{5.69}$$

and:

$$0 \leq \mu_{f}^{-1} \left\| \frac{\partial y_{NN}}{\partial f} \right\|^2 - \mu_{f}^{-1} \left\| \frac{\partial y_{NN}}{\partial f} \right\|^4 \left(\mu_{f} + \left\| \frac{\partial y_{NN}}{\partial f} \right\|^2\right)^{-1} \tag{5.70}$$

Hence, in order to have a negative time difference for $V(k)$, we must have:

$$0 \leq \frac{1}{2} - \mu_{\overline{\sigma}}^{-1} \left\| \frac{\partial y_{NN}}{\partial \overline{\sigma}} \right\|^2 + \mu_{\overline{\sigma}}^{-1} \left\| \frac{\partial y_{NN}}{\partial \overline{\sigma}} \right\|^4 \left(\mu_{\overline{\sigma}} + \left\| \frac{\partial y_{NN}}{\partial \overline{\sigma}} \right\|^2 \right)^{-1} \tag{5.71}$$

and:

$$0 \leq \frac{1}{2} - \mu_{\underline{\sigma}}^{-1} \left\| \frac{\partial y_{NN}}{\partial \underline{\sigma}} \right\|^2 + \mu_{\underline{\sigma}}^{-1} \left\| \frac{\partial y_{NN}}{\partial \underline{\sigma}} \right\|^4 \left(\mu_{\underline{\sigma}} + \left\| \frac{\partial y_{NN}}{\partial \underline{\sigma}} \right\|^2 \right)^{-1} \tag{5.72}$$

and:

$$0 \leq \frac{1}{2} - \mu_{c}^{-1} \left\| \frac{\partial y_{NN}}{\partial c} \right\|^2 + \mu_{c}^{-1} \left\| \frac{\partial y_{NN}}{\partial c} \right\|^4 \left(\mu_{c} + \left\| \frac{\partial y_{NN}}{\partial c} \right\|^2 \right)^{-1} \tag{5.73}$$

and:

$$0 \leq \frac{1}{2} - \mu_{f}^{-1} \left\| \frac{\partial y_{NN}}{\partial f} \right\|^2 + \mu_{f}^{-1} \left\| \frac{\partial y_{NN}}{\partial f} \right\|^4 \left(\mu_{f} + \left\| \frac{\partial y_{NN}}{\partial f} \right\|^2 \right)^{-1} \tag{5.74}$$

Equation (5.71) can be further simplified as:

$$0 \leq \left(\frac{1}{2} - \mu_{\overline{\sigma}}^{-1} \left\| \frac{\partial y_{NN}}{\partial \overline{\sigma}} \right\|^2 \right) \left(\mu_{\overline{\sigma}} + \left\| \frac{\partial y_{NN}}{\partial \overline{\sigma}} \right\|^2 \right) + \mu_{\overline{\sigma}}^{-1} \left\| \frac{\partial y_{NN}}{\partial \overline{\sigma}} \right\|^4 \tag{5.75}$$

Thus,

$$0 \leq \frac{1}{2}\mu_{\overline{\sigma}} - \frac{1}{2} \left\| \frac{\partial y_{NN}}{\partial \overline{\sigma}} \right\|^2 \tag{5.76}$$

and finally the following constraint is obtained for $\mu_{\overline{\sigma}}$:

$$\left\| \frac{\partial y_{NN}}{\partial \overline{\sigma}} \right\|^2 \leq \mu_{\overline{\sigma}} \tag{5.77}$$

Using a similar analysis the following constraints are obtained for $\mu_{\underline{\sigma}}, \mu_c$, and μ_f:

$$\left\| \frac{\partial y_{NN}}{\partial \underline{\sigma}} \right\|^2 \leq \mu_{\underline{\sigma}}, \left\| \frac{\partial y_{NN}}{\partial c} \right\|^2 \leq \mu_c, \left\| \frac{\partial y_{NN}}{\partial f} \right\|^2 \leq \mu_f \tag{5.78}$$

5.5 FURTHER READING

Interested readers may refer to [1, 8] for more information and a detailed discussion of existing nonlinear programming techniques. Different variants of GD as applied to ANNs may be found in Ref. [3].

5.6 CONCLUSION

This chapter introduces the basic notion and mathematics of GD-based training methods including its benefits and challenges. The stability analysis of these methods are also considered using appropriate Lyapunov functions. GD-based training methods are one of the most commonly used optimization algorithms to estimate the parameters of T2FNNs. However, since these methods are based on the first-order gradient, there is always the possibility of entrapment in a local minima. The second-order gradient methods, such as LM, may lessen the probability of entrapment in local minima. However, they never fully solve the aforementioned problem. The calculation of a gradient with respect to the parameters of T2FNNs, especially the parameters of the antecedent part, is also difficult and does not have a closed form. In addition, since these rules need some matrix manipulations and partial derivatives, they are not easy to implement in real-time systems.

REFERENCES

[1] D.P. Bertsekas, Nonlinear Programming, Athena Scientific, Belmont, MA, 1999.

[2] T. Cong Chen, D. Jian Han, F. Au, L. Tham, Acceleration of Levenberg-Marquardt training of neural networks with variable decay rate, in: Neural Networks, 2003. Proceedings of the International Joint Conference on, vol. 3, 2003, pp. 1873-1878.

[3] A. Cochocki, R. Unbehauen, Neural Networks for Optimization and Signal Processing, John Wiley & Sons, Inc., New York, 1993.

[4] C.-H. Lee, C.-C. Teng, Identification and control of dynamic systems using recurrent fuzzy neural networks, IEEE Trans. Fuzzy Syst. 8 (4) (2000) 349-366.

[5] M.A. Shoorehdeli, M. Teshnehlab, A.K. Sedigh, M.A. Khanesar, Identification using ANFIS with intelligent hybrid stable learning algorithm approaches and stability analysis of training methods, Appl. Soft Comput. 9 (2) (2009) 833-850.

[6] M.A. Shoorehdeli, M. Teshnehlab, A.K. Sedigh, Training ANFIS as an identifier with intelligent hybrid stable learning algorithm based on particle swarm optimization and extended Kalman filter, Fuzzy Sets Syst. 160 (7) (2009) 922-948.
[7] K.J. Åström, B. Wittenmark, Adaptive control, Courier Corporation, Mineola, NY, 2013.
[8] M.S. Bazaraa, H.D. Sherali, C.M. Shetty, Nonlinear Programming: Theory and Algorithms, John Wiley & Sons, New York, 2013.

CHAPTER 6

Extended Kalman Filter Algorithm for the Tuning of Type-2 Fuzzy Neural Networks

Contents

Abstract

In this chapter, the EKF algorithm is used to optimize the parameters of T2FNNs. The basic version of KF is an optimal linear estimator where system is linear and is subject to white uncorrelated noise. However, it is possible to use Taylor expansion to extend its applications to nonlinear cases. Finally, the decoupled version of the EKF is also discussed, which is computationally more efficient than EKF to tune the parameters of T2FNNs.

Keywords

Extended Kalman filter, Kalman filter, Estimation, Type-2 fuzzy neural networks

6.1 INTRODUCTION

KF uses a series of measurements observed over time containing noise, and produces state estimations that tend to be more precise than those based on a single measurement alone. This method is an efficient algorithm that needs to store only the last estimation of the states and the covariance matrix of the error to estimate the current value of the states of the system. Because of the algorithm's recursive nature, it can run in real time using only the present

Fuzzy Neural Networks for Real Time Control Applications
http://dx.doi.org/10.1016/B978-0-12-802687-8.00006-2

input measurements and the previously calculated states and its covariance matrix; no additional past information is required. Although this algorithm was originated from linear dynamic system theory, where it serves as an observer for the unmeasured states of the system, it has been successfully extended to nonlinear estimation applications using Taylor expansion of the nonlinear functions.

6.2 DISCRETE TIME KALMAN FILTER

Consider a linear discrete time dynamic system in state space form as follows:

$$\begin{aligned} \theta_{k+1} &= F_{k+1,k}\theta_k + G_{k+1,k}u_k + w_k \\ y_k &= H_k\theta_k + v_k \end{aligned} \tag{6.1}$$

where $\theta_k \in R^n$ is the kth sample of the state vector of the system, $u_k \in R^m$ is the kth input of the system, $w \in R^n$ is the process noise, v is the measurement noise, $y \in R^l$ is the measured output of the system, and $F_{k,k+1}$, $G_{k,k+1}$, and H_k are the state matrices of the system with appropriate dimensions. The noise processes w_k and v_k are considered to be white, uncorrelated and zero-mean with the following properties:

$$\begin{aligned} E[w_k] &= 0 \\ E[v_k] &= 0 \\ E[w_k w_j^{\mathrm{T}}] &= Q_k\delta_{j-k} \\ E[v_k v_j^{\mathrm{T}}] &= R_k\delta_{j-k} \\ E[v_k w_j^{\mathrm{T}}] &= 0 \end{aligned}$$

We have the following theorems for optimal estimates of the states of the system:

Theorem 6.1 (Conditional Mean Estimator [1–3]). *If the stochastic processes $\{\theta_k\}$ and $\{y_k\}$ are jointly Gaussian, then the optimum estimate $\hat{\theta}_k$ minimizes the mean-square error J_k, which is defined as follows:*

$$J_k = E[(\theta_k - \hat{\theta}_k)^2] \tag{6.2}$$

The following is the conditional mean estimator:

$$\hat{\theta}_k = E[\theta_k | y_1, y_2, \ldots, y_k]. \tag{6.3}$$

Theorem 6.2 (Principle of Orthogonality [1–3]). *Let the stochastic processes* $\{\theta_k\}$ *and* $\{y_k\}$ *be of zero means, i.e.,*

$$E[\theta_k] = E[y_k] = 0, \quad \forall k \tag{6.4}$$

If either:

(i) *the stochastic process* $\{\theta_k\}$ *and* $\{y_k\}$ *are jointly Gaussian; or*

(ii) *if the optimal state estimate* $\hat{\theta}_k$*, which is the solution to the mean-square cost function is restricted to be a linear function of the observed variables,*

then the optimum state estimate $\hat{\theta}_k$*, given the observed variables* $y_1, y_2, \ldots, y_k$*, is the orthogonal projection of* θ_k *on the space spanned by these observations.*

The variable $\hat{\theta}_k^-$ is defined as a priori estimate of the states of the system at kth sample, based on the measurements up to $(k-1)$th sample i.e. $\hat{\theta}_k^- = E[\theta_k | y_{k-1}]$.

$$\begin{aligned}\hat{\theta}_k^- &= E[\theta_k | y_{k-1}] \\ &= E[F_{k,k-1}\theta_{k-1} + G_{k,k-1}u_{k-1} + w_{k-1} | y_{k-1}]\end{aligned} \tag{6.5}$$

Since w is a white noise that is uncorrelated from the output, we have the following:

$$\hat{\theta}_k^- = E[F_{k,k-1}\theta_{k-1} + G_{k,k-1}u_{k-1} | y_{k-1}] \tag{6.6}$$

The posteriori estimate of the states of the system at the $(k-1)$th sample using the measurement up to the $(k-1)$th sample is called $\hat{\theta}_{k-1}$ and is defined as $\hat{\theta}_{k-1} = E[\theta_{k-1} | y_{k-1}]$. Therefore, it is possible to use $\hat{\theta}_{k-1}$ to estimate a priori estimate for θ_k using the dynamic equation of the system as follows:

$$\hat{\theta}_k^- = F_{k,k-1}\hat{\theta}_{k-1} + G_{k,k-1}u_{k-1} \tag{6.7}$$

Furthermore, we have the following equations:

$$\begin{aligned}F_{k,k-1}\theta_{k-1} &= F_{k,k-1}\hat{\theta}_{k-1}^- + F_{k,k-1}(\theta_{k-1} - \hat{\theta}_{k-1}^-) \\ &= F_{k,k-1}\hat{\theta}_{k-1}^- + F_{k,k-1}e_{k-1}^-\end{aligned} \tag{6.8}$$

where e_{k-1}^- is the error of a priori estimate, which is defined as follows:

$$e_{k-1}^- = \theta_{k-1} - \theta_{k-1}^- \tag{6.9}$$

The principle of orthogonality requires that e_{k-1}^- and y_{k-1} are orthogonal, i.e.,

$$E[e_{k-1}^- | y_{k-1}] = 0 \tag{6.10}$$

Moreover, using (6.7), we have the following equation for the error between a priori estimate $\hat{\theta}_k^-$ and θ_k:

$$\begin{aligned} e_k^- &= \theta_k - \hat{\theta}_k^- \\ &= F_{k,k-1}\theta_{k-1} + G_{k,k-1}u_{k-1} + w_{k-1} - F_{k,k-1}\hat{\theta}_{k-1} - G_{k,k-1}u_{k-1} \\ &= F_{k,k-1}e_{k-1} + w_{k-1} \end{aligned} \tag{6.11}$$

where e_{k-1} is defined as $e_{k-1} = \theta_{k-1} - \hat{\theta}_{k-1}$. A priori covariance matrix is defined as follows:

$$P_k^- = E[e_k^- (e_k^-)^{\mathrm{T}}] \tag{6.12}$$

Using (6.11), we have the following equation for P_k^-:

$$P_k^- = E[e_k^- (e_k^-)^{\mathrm{T}}] \tag{6.13}$$

$$= E[(F_{k,k-1}e_{k-1} + w_{k-1})(F_{k,k-1}e_{k-1} + w_{k-1})^{\mathrm{T}}] \tag{6.14}$$

Since e_{k-1} and w_{k-1} are uncorrelated, we have:

$$P_k^- = F_{k,k-1}E[e_{k-1}(e_{k-1})^{\mathrm{T}}]F_{k,k-1} + E[w_{k-1}w_{k-1}^{\mathrm{T}}] \tag{6.15}$$

A posteriori covariance matrix at $(k-1)$th sample (P_{k-1}) is defined as follows:

$$P_{k-1} = E[e_{k-1}e_{k-1}^{\mathrm{T}}] \tag{6.16}$$

Using (6.16) and the properties of w as in (6.2), we have:

$$P_k^- = F_{k,k-1}P_{k-1}F_{k,k-1} + Q_{k-1} \tag{6.17}$$

A posteriori estimate of θ_k is considered to be a linear combination of θ_k^- and the current measured output of the system y_k as follows:

$$\hat{\theta}_k = L_k\theta_k^- + K_k y_k \tag{6.18}$$

From the principles of orthogonality, in order to have the optimal estimate, the error of the estimation $\theta_k - \hat{\theta}_k$ and y_k are required to be orthogonal and hence the following equation holds:

$$\begin{aligned} 0 &= E[\theta_k - \hat{\theta}_k | y_k] \\ &= E[\theta_k - L_k\hat{\theta}_k^- - K_kH_k\theta_k - K_kv_k | y_k] \\ &= E[\theta_k - L_k\theta_k + I_ke_k^- - K_kH_k\theta_k | y_k] \\ &= (I - L_k - K_kH_k)E[\theta_k | y_k] \end{aligned} \tag{6.19}$$

where:

$$e_k^- = \theta_k - \hat{\theta}_k^- \tag{6.20}$$

Since we have $E[\theta_k | y_k] \neq 0$, from (6.19), it is required that:

$$L_k = I - K_kH_k \tag{6.21}$$

Using L_k as in (6.21) and the estimator equation of (6.18) and the dynamic equation of the system (6.1), we have the following:

$$\begin{aligned} e_k &= \theta_k - \hat{\theta}_k \\ &= \theta_k - (1 - K_kH_k)\hat{\theta}_k^- - K_ky_k \\ &= F_{k,k-1}\theta_{k-1} + G_{k,k-1}u_{k-1} + w_{k-1} - (I - K_kH_k)\hat{\theta}_k^- - K_ky_k \\ &= F_{k,k-1}\hat{\theta}_{k-1} + G_{k,k-1}u_{k-1} + F_{k,k-1}e_{k-1} + w_{k-1} - \hat{\theta}_k^- \\ &\quad + K_kH_ke_k^- - K_kv_k \end{aligned} \tag{6.22}$$

Using (6.7), we have:

$$\begin{aligned} e_k &= F_{k,k-1}\hat{\theta}_{k-1} + G_{k,k-1}u_{k-1} + F_{k,k-1}e_{k-1} + w_{k-1} \\ &\quad - (F_{k,k-1}\hat{\theta}_{k-1} + G_{k,k-1}u_{k-1}) + K_kH_ke_k^- - K_kv_k \\ &= F_{k,k-1}e_{k-1} + w_{k-1} - K_kH_ke_k^- - K_kv_k \end{aligned} \tag{6.23}$$

Using (6.11) and (6.23), we have:

$$\begin{aligned} e_k &= F_{k,k-1}e_{k-1} + w_{k-1} - K_kH_k(F_{k,k-1}e_{k-1} + w_{k-1}) \\ &= (I - K_kH_k)F_{k,k-1}e_{k-1} + (I - K_kH_k)w_{k-1} - K_kv_k \end{aligned} \tag{6.24}$$

Therefore, we have the following equations for a posteriori covariance matrix:

$$\begin{aligned} P_k &= E[e_k e_k^{\mathrm{T}}] \\ &= (I - K_k H_k) F_{k,k-1} E[e_{k-1} e_{k-1}^{\mathrm{T}}] F_{k,k-1}^{\mathrm{T}} (I - K_k H_k)^{\mathrm{T}} \\ &\quad + (I - K_k H_k) E[w_{k-1} w_{k-1}^{\mathrm{T}}] (I - K_k H_k)^{\mathrm{T}} \\ &\quad + K_k E[v_k v_k^{\mathrm{T}}] K_k^{\mathrm{T}} \\ &= (I - K_k H_k) F_{k,k-1} P_{k-1} F_{k,k-1}^{\mathrm{T}} (I - K_k H_k)^{\mathrm{T}} \\ &\quad + (I - K_k H_k) Q_{k-1} (I - K_k H_k)^{\mathrm{T}} \\ &\quad + K_k R_k K_k^{\mathrm{T}} \end{aligned} \tag{6.25}$$

Furthermore, the state estimation error $\theta_k - \hat{\theta}_k$ and the output estimation error are uncorrelated, so that:

$$\begin{aligned} 0 &= E[(\theta_k - \hat{\theta}_k)(y_k - H_k \hat{\theta}_k^{-})^{\mathrm{T}}] \\ &= E[(\theta_k - (I - K_k H_k)\hat{\theta}_k^{-} - K_k y_k)(y_k - H_k \hat{\theta}_k^{\mathrm{T}})^{\mathrm{T}}] \\ &= E[(e_k^{-} - K_k H_k e_k^{-} - K_k v_k)(H_k e_k^{-} + v_k)^{\mathrm{T}}] \\ &= (I - K_k H_k) E[e_k^{-} (e_k^{-})^{\mathrm{T}}] - K_k E[v_k v_k^{\mathrm{T}}] \end{aligned} \tag{6.26}$$

Considering (6.2) and (6.13), from (6.26) we have:

$$(I - K_k H_k) P_k^{-} H_k^{\mathrm{T}} - K_k R_k = 0 \tag{6.27}$$

which can be rewritten as:

$$K_k (H_k P_k^{-} H_k^{\mathrm{T}} + R_k) = P_k^{-} H_k^{\mathrm{T}} \tag{6.28}$$

Hence, K_k, the Kalman gain, is obtained as follows:

$$K_k = P_k^{-} H_k^{\mathrm{T}} (H_k P_k^{-} H_k^{\mathrm{T}} + R_k)^{-1} \tag{6.29}$$

By applying (6.29) to (6.25) we have the following equation for the covariance matrix:

$$\begin{aligned} P_k &= (I - K_kH_k)P_k^- + K_kH_kP_k^-H_k^TK_k^T - P_k^-H_k^TK_k^T + K_kP_kK_k^T \\ &= (I - K_kH_k)P_k^- + K_k(H_kP_k^-K_k^T + R_k)K_k^T - P_k^-H_k^TK_k^T \\ &= (I - K_kH_k)P_k^- + P_k^-H_k^TK_k^T - P_k^-H_k^TK_k^T \\ &= (I - K_kH_k)P_k^- \end{aligned} \tag{6.30}$$

Table 6.1 summarizes the KF which is required to update the estimated values of the states of the system (6.1). It can be shown that if w_k and v_k are white, zero-mean and uncorrelated, then KF is an optimal linear estimator in terms of a weighted mean squared error cost function [4]. In other words, it is possible to have a nonlinear filter that performs better than KF, but no linear filter can act better than it. In Ref. [4], it is stated that it is not necessary to have Gaussian noise as an optimality requirement for KF.

Table 6.1 Summary of the KF

Let the state space model of the system be:

$$\begin{aligned} \theta_{k+1} &= F_{k+1,k}\theta_k + G_{k+1,k}u_k + w_k \\ y_k &= H_k\theta_k + v_k \end{aligned}$$

where w_k and v_k are uncorrelated, zero-mean, and Gaussian noise processes with their covariance matrix equal to Q_k and R_k, respectively.
The initial conditions for the estimator are considered to be:

$$\begin{aligned} \hat{\theta}_0 &= E[\theta_0] \\ P_0 &= E[(\theta_0 - E[\theta_0])(\theta_0 - E[\theta_0])^T] \end{aligned}$$

A priori estimate: a priori estimate of θ_k is obtained from a past estimate and the dynamics of the system as:

$$\hat{\theta}_k^- = F_{k,k-1}\hat{\theta}_{k-1} + G_{k,k-1}u_{k-1}$$

A priori estimate of covariance matrix:

$$P_k^- = F_{k,k-1}P_{k-1}F_{k,k-1}^T + Q_{k,k-1}$$

The Kalman gain:

$$K_k = P_k^-H_k^T[H_kP_k^-H_k^T + R_k]^{-1}$$

State estimate update:

$$\hat{\theta}_k = \theta_k^- + K_k(y_k - H_k\theta_k^-)$$

Error covariance update

$$P_k = (I - K_kH_k)P_k^-$$

6.3 SQUARE-ROOT FILTERING

Although KF is an efficient and optimal linear estimator in the present of noise, it may suffer from numerical difficulties, which happens mostly as a result of finite length of words used by a digital computer [4]. One of the main problems, as discussed in Ref. [4], is that it is possible the covariance matrix becomes ill condition as a result of large condition number value. A possible solution to this problem may be using the square root of the covariance matrix. The square root of P_k is called S_k. The condition number of S_k is proven to be equal to the square root of the condition number of P_k, i.e.:

$$\sigma(S_k) = \sqrt{\sigma(P_k)} \tag{6.31}$$

where $\sigma(.)$ indicates the condition number of its corresponding argument. This modification may solve some of the problems caused by a large number of P_k. The square-root filtering algorithm, which is discussed here is based on Potter's algorithm [5, 6], which was further modified to deal with vector measurements [7].

It is considered that R_k is a diagonal matrix as $R_k = \text{diag}(R_{1k}, \ldots, R_{ik}, \ldots, R_{lk})$. Therefore, R_{ik} is the variance of measurement noise on ith measurement. Furthermore, H_{ik} is defined as the ith row of the matrix H_k. Table 6.2 summarizes this algorithm. Although root-square type of KF is numerically more robust, its computational cost is much higher than the original version of KF.

6.4 EXTENDED KALMAN FILTER ALGORITHM

It is possible to extend the original version of KF to nonlinear dynamic systems using first-order Taylor expansion of nonlinear functions. Table 6.3 summarizes this algorithm [4].

6.5 EXTENDED KALMAN FILTER TRAINING OF TYPE-2 FUZZY NEURAL NETWORKS

As a powerful parameter estimator, EKF is one of the most successful training algorithms applied to NNs [8] and FNN structure [9–11]. In order to discuss the EKF training method for T2FNN structure, a simple type-2 structure

Table 6.2 Summary of the root square type of KF [4]

1. Let θ_k^- and S_k^- be the a priori state estimate and the a priori covariance square root, respectively.
   Initialize:

$$\hat{\theta}_{0k} = \hat{\theta}_k^-$$

$$S_{0k}^+ = S_k^-$$

2. For $i = 1, \ldots, r$ (where r is the number of measurements. Perform the following steps:
   (a) Let H_{ik} be the ith row of H_k, and y_{ik} be the ith measurement, and R_{ik} be the variance of the ith measurement.
   (b) Perform the following to find the square root of the variance after the ith measurement has been processed:

$$\phi_i = S_{i-1,k}^{+T} H_{ik}^{\mathrm{T}}$$

$$a_i = \frac{1}{\phi_i^{\mathrm{T}}\phi_i + R_{ik}}$$

$$\gamma_i = \frac{1}{1 \pm \sqrt{a_i R_{ik}}}$$

$$S_{ik}^+ = S_{i-1,k}^+ (I - a_i \gamma_i \phi_i \phi_i^{\mathrm{T}})$$

   (c) Compute the Kalman gain for the ith measurement as:

$$K_{ik} = a_i S_{ik}^+ \phi_i$$

   (d) Update the estimation of the states due to ith measurement as follows:

$$\hat{\theta}_{ik} = \hat{\theta}_{i-1,k} + K_{ik}(y_{ik} - K_{ik}\hat{\theta}_{i-1,k})$$

3. Set the a posteriori covariance square root and the a posteriori state estimate as:

$$S_k^+ = S_{rk}^+$$

$$\hat{\theta}_k = \hat{\theta}_{rk}$$

is investigated in this section. Consider a T2FNN with two inputs and I number of MFs for the first input and J number of MFs for the second input. The upper MFs for the first input x_1 are considered to be $\overline{\mu}_{11}(x_1)$, $\overline{\mu}_{12}(x_1), \ldots,$ and $\overline{\mu}_{1I}(x_1)$, and its corresponding lower MFs are considered to be $\underline{\mu}_{11}(x_1)$, $\underline{\mu}_{12}(x_1), \ldots$ and $\underline{\mu}_{1I}(x_1)$. The upper MFs for the second input (x_2) are considered to be $\overline{\mu}_{21}(x_2)$, $\overline{\mu}_{22}(x_2), \ldots,$ and $\overline{\mu}_{2J}(x_2)$, and its

Table 6.3 Summary of the EKF algorithm

1. Let the nonlinear state space model of the system be:

$$\begin{aligned}\theta_k &= f_{k-1}(\theta_{k-1}, u_{k-1}, w_{k-1})\\ y_k &= h_k(\theta_k, v_k)\end{aligned}$$

where w_k and v_k are uncorrelated, zero–mean and Gaussian noise processes with following properties:

$$\begin{aligned}E[w_k w_k^{\mathrm{T}}] &= Q_k\\ E[v_k v_k^{\mathrm{T}}] &= R_k\end{aligned}$$

2. The initial conditions for the estimator are considered to be as:

$$\begin{aligned}\hat{\theta}_0 &= E[\theta_0]\\ P_0 &= E[(\theta_0 - E[\theta_0])(\theta_0 - E[\theta_0])^{\mathrm{T}}]\end{aligned}$$

3. For $k = 1, 2, \ldots$, perform the following steps:
(a) Compute the following partial derivatives:

$$\begin{aligned}F_{k-1} &= \frac{\partial f_{k-1}}{\partial \theta}|_{\hat{\theta}_{k-1}}\\ L_{k-1} &= \frac{\partial f_{k-1}}{\partial w}|_{\hat{\theta}_{k-1}}\end{aligned}$$

(b) A priori estimate: a priori estimate of θ_k is obtained as follows:

$$\hat{\theta}_k^- = f_{k-1}(\hat{\theta}_{k-1}, u_{k-1}, 0)$$

(c) A priori estimate for covariance matrix:

$$P_k^- = F_{k-1}P_{k-1}F_{k-1}^{\mathrm{T}} + L_{k-1}Q_{k-1}L_{k-1}^{\mathrm{T}}$$

(d) Compute the following partial derivatives:

$$\begin{aligned}H_k &= \frac{\partial h_k}{\partial \theta}|_{\hat{\theta}_k^-}\\ M_k &= \frac{\partial h_k}{\partial v}|_{\hat{\theta}_k^-}\end{aligned}$$

(e) Update the Kalman gain as:

$$K_k = P_k^- H_k^{\mathrm{T}}[H_k P_k^- H_k^{\mathrm{T}} + M_k R_k M_k^{\mathrm{T}}]^{-1}$$

(f) Update the state estimate as:

$$\hat{\theta}_k = \theta_k^- + K_k(y_k - h_k(\theta_k^-, 0))$$

(g) Update the error covariance matrix as:

$$P_k = (I - K_k H_k)P_k^-$$

corresponding lower MFs are considered to be $\underline{\mu}_{21}(x_2)$, $\underline{\mu}_{22}(x_2), \ldots$, and $\underline{\mu}_{2J}(x_2)$. The terms $\underline{\mu}_{1i}(x_1)$, $\overline{\mu}_{1i}(x_1)$, $\underline{\mu}_{2j}(x_2)$, and $\overline{\mu}_{2j}(x_2)$ are considered to be defined as follows:

$$\underline{\mu}_{1i}(x_1) = \exp\left(-\left(\frac{x_1 - c_{1i}}{\underline{\sigma}_{1i}}\right)^2\right) \tag{6.32}$$

$$\overline{\mu}_{1i}(x_1) = \exp\left(-\left(\frac{x_1 - c_{1i}}{\overline{\sigma}_{1i}}\right)^2\right) \tag{6.33}$$

$$\underline{\mu}_{2j}(x_2) = \exp\left(-\left(\frac{x_2 - c_{2j}}{\underline{\sigma}_{2j}}\right)^2\right) \tag{6.34}$$

$$\overline{\mu}_{2j}(x_2) = \exp\left(-\left(\frac{x_2 - c_{2j}}{\overline{\sigma}_{2j}}\right)^2\right) \tag{6.35}$$

The upper and the lower rules of the system are derived as follows:

$$\begin{aligned}
\underline{w}_1 &= \underline{\mu}_{11}(x_1)\underline{\mu}_{21}(x_2), \ldots, \underline{w}_{1\times J} = \underline{\mu}_{11}(x_1)\underline{\mu}_{2J}(x_2) \\
\underline{w}_{1\times J+1} &= \underline{\mu}_{12}(x_1)\underline{\mu}_{21}(x_2), \ldots, \underline{w}_{2\times J} = \underline{\mu}_{12}(x_1)\underline{\mu}_{2J}(x_2) \\
&\vdots \\
\underline{w}_{(I-1)\times J+1} &= \underline{\mu}_{1I}(x_1)\underline{\mu}_{21}(x_2), \ldots, \underline{w}_{I\times J} = \underline{\mu}_{1I}(x_1)\underline{\mu}_{2J}(x_2)
\end{aligned} \tag{6.36}$$

and

$$\begin{aligned}
\overline{w}_1 &= \overline{\mu}_{11}(x_1)\overline{\mu}_{21}(x_2), \ldots, \overline{w}_{1\times J} = \overline{\mu}_{11}(x_1)\overline{\mu}_{2J}(x_2) \\
\overline{w}_{1\times J+1} &= \overline{\mu}_{12}(x_1)\overline{\mu}_{21}(x_2), \ldots, \overline{w}_{2\times J} = \overline{\mu}_{12}(x_1)\overline{\mu}_{2J}(x_2) \\
&\vdots \\
\overline{w}_{(I-1)\times J+1} &= \overline{\mu}_{1I}(x_1)\overline{\mu}_{21}(x_2), \ldots, \overline{w}_{I\times J} = \overline{\mu}_{1I}(x_1)\overline{\mu}_{2J}(x_2)
\end{aligned} \tag{6.37}$$

The output of the T2FNN is obtained as follows:

$$y_N = q \sum_{r=1}^{N=I\times J} f_r \tilde{\underline{w}}_r + (1-q) \sum_{r=1}^{N=I\times J} f_r \tilde{\overline{w}}_r \tag{6.38}$$

where:

$$\tilde{\underline{w}}_r = \frac{\underline{w}_r}{\sum_{l=1}^{N} \underline{w}_l}, \quad \tilde{\overline{w}}_r = \frac{\overline{w}_r}{\sum_{l=1}^{N} \overline{w}_l} \tag{6.39}$$

and:

$$f_r = a_{ri}x_i + b_r, \quad \forall r = 1, 2, \ldots, N, \quad \text{and} \quad i = 1, 2, \ldots, n \tag{6.40}$$

In order to apply EKF to T2FNN, all unknown parameters of T2FLS must be gathered in a vector. The vector of unknown parameters has the following form:

$$\begin{aligned}\theta = [&c_{11}, \ldots, c_{1i}, \ldots, c_{1J}, c_{21}, \ldots, c_{2j}, \ldots, c_{2J}, \underline{\sigma}_{11}, \ldots, \underline{\sigma}_{1i}, \ldots, \underline{\sigma}_{1J}, \overline{\sigma}_{11}, \\ &\ldots, \overline{\sigma}_{1i}, \ldots, \overline{\sigma}_{1J}, \underline{\sigma}_{21}, \ldots, \underline{\sigma}_{2j}, \ldots, \underline{\sigma}_{2J}, \overline{\sigma}_{21}, \ldots, \overline{\sigma}_{2j}, \ldots, \overline{\sigma}_{2J}, a_{11}, \ldots, \\ &a_{1n}, \ldots, a_{l1}, \ldots, a_{ln}, \ldots, a_{N1}, \ldots, a_{Nn}, b_1, \ldots, b_N]\end{aligned} \tag{6.41}$$

The partial derivative of the output of the fuzzy system with respect to θ in each sample k must be calculated. These calculations have been previously discussed in Chapter 5. Having calculated these partial derivatives, $H_k = \frac{\partial y_N}{\partial \theta}$ is derived. Since the parameters of T2FNN are time-invariant, the process equation is derived as follows:

$$\theta_{k+1} = \theta_k \tag{6.42}$$

This equation implies that $f_{k-1}(\theta_{k-1}) = \theta_{k-1}$ and F_{k-1} is equal to the identity matrix. The rest of the algorithm is as that in Table 6.3.

However, since all the adaptive parameters of T2FNN are gathered in a vector, the computational cost of EKF is of the order of AB^2 where A is the dimension of the output of dynamical system and B is the number of the parameters. The total number of parameters for a T2FNN structure depends on the number of inputs and the number of MFs considered for each input. For a T2FNN that has n inputs, c Gaussian type-2 MFs with uncertain σ value for each input, N number of rules, and an output, the total number of adaptive parameters is $3c + (n+1)N$. The total number of the rules N is a large number and is multiplied by the number of MF of inputs. The computational cost of EKF when applied to T2FNN will be $9c^2 + (n+1)^2N^2 + 6c(n+1)N$, which is a very big number. Furthermore, the size of covariance matrix will be equal to $9c^2 + (n+1)^2N^2 + 6c(n+1)N$, which occupies a large amount of memory. In order to reduce the computational cost of EKF, its decoupled version is mostly used in the training of a complex structure such as NNs [3, 8], FNNs [10] and T2FNNs [11].

6.6 DECOUPLED EXTENDED KALMAN FILTER

In order to reduce the computational cost of EKF and reduce its memory consumption, most researchers prefer to use it in a decoupled form. The DEKF is based on the assumption that certain parameters interact with each other only at a second-order level. The DEKF has been previously used in Refs. [8, 11]. For the T2FNN considered in this chapter, there are three adaptive parameters for each type-2 MF. The parameters of the consequent part are put in one vector, and they make one group of parameters. In this way, four groups of parameters are considered:

$$\theta^1 = [c_{11}, \ldots, c_{1i}, \ldots, c_{1I}, c_{21}, \ldots, c_{2j}, \ldots, c_{2J}] \tag{6.43}$$

$$\theta^2 = [\underline{\sigma}_{11}, \ldots, \underline{\sigma}_{1i}, \ldots, \underline{\sigma}_{1I}, \underline{\sigma}_{21}, \ldots, \underline{\sigma}_{2j}, \ldots, \underline{\sigma}_{2J}] \tag{6.44}$$

$$\theta^3 = [\overline{\sigma}_{11}, \ldots, \overline{\sigma}_{1i}, \ldots, \overline{\sigma}_{1I}, \overline{\sigma}_{21}, \ldots, \overline{\sigma}_{2j}, \ldots, \overline{\sigma}_{2J}] \tag{6.45}$$

$$\theta^4 = [a_{11}, \ldots, a_{1n}, \ldots, a_{l1}, \ldots, a_{ln}, \ldots, a_{N1}, \ldots, a_{Nn}, b_1, \ldots, b_N] \tag{6.46}$$

Each group of the parameters has its own covariance matrix and Kalman gain. The computational cost of estimation is greatly reduced and it is of the order of $3c^2 + (n+1)^2N^2$. The ratio of computational cost of decoupled EKF to standard EKF is shown as follows:

$$\frac{\text{Decoupled EKF computational cost}}{\text{Standard EKF computational cost}} = \frac{3c^2 + (n+1)^2r^2}{9c^2 + (n+1)^2N^2 + 6c(n+1)N} \tag{6.47}$$

The memory consumption is also reduced with the same order.

6.7 CONCLUSION

In this chapter, after a short introduction to KF and its variants, it is extended to estimate the parameters of nonlinear systems, particularly T2FNNs. The extended version of KF is called EKF, which is a powerful tool to optimize the parameters of T2FNNs. Although KF can be shown to be an optimal linear estimator for linear systems subject to process and measurement white noise, its extended version is a suboptimal method because it uses the first-order Taylor expansion of nonlinear functions. Furthermore, because of the large number of adaptive parameters in T2FNN structure, EKF is shown be computationally very demanding with huge memory consumption. In order to deal with this problem and reduce the size of the covariance matrix, it is possible to assume that certain parameters interact with each other only at a second-order level. In this way, it is possible to consider small groups for the parameters, which greatly reduces the computational cost and memory consumption of the algorithm.

REFERENCES

[1] R.E. Kalman, A new approach to linear filtering and prediction problems, J. Fluids Eng. 82 (1) (1960) 35-45.
[2] H.L. Van Trees, Detection, Estimation, and Modulation Theory, John Wiley Sons, New York, 2004.
[3] S.S. Haykin, Kalman Filtering and Neural Networks, Wiley Online Library, New York, 2001.
[4] D. Simon, Optimal State Estimation: Kalman, H infinity, and Nonlinear Approaches, John Wiley & Sons, New York, 2006.
[5] R.H. Battin, Astronautical guidance, McGraw-Hill, 1964.
[6] P. Kaminski, A.E. Bryson, S. Schmidt, Discrete square root filtering: A survey of current techniques, IEEE Trans. Automat. Control 16 (6) (1971) 727-736.
[7] A. Andrews, A square root formulation of the Kalman covariance equations, AIAA J. 6 (6) (1968) 1165-1166.
[8] D. Simon, Training radial basis neural networks with the extended Kalman filter, Neurocomputing 48 (1-4) (2002) 455-475.
[9] M.A. Shoorehdeli, M. Teshnehlab, A.K. Sedigh, Training ANFIS as an identifier with intelligent hybrid stable learning algorithm based on particle swarm optimization and extended Kalman filter, Fuzzy Sets Syst. 160 (7) (2009) 922-948.
[10] D. Simon, Training fuzzy systems with the extended Kalman filter, Fuzzy Sets Syst. 132 (2) (2002) 189-199.
[11] M. Khanesar, E. Kayacan, M. Teshnehlab, O. Kaynak, Extended Kalman filter based learning algorithm for type-2 fuzzy logic systems and its experimental evaluation, IEEE Trans. Indust. Electron. 59 (11) (2012) 4443-4455.

CHAPTER 7

Sliding Mode Control Theory-Based Parameter Adaptation Rules for Fuzzy Neural Networks

Contents

Abstract

In this chapter, in order to deal with nonlinearities, lack of modeling several uncertainties, and noise in both identification and control problems, SMC theory-based learning algorithms are designed to tune both the premise and consequent parts of T2FNNs. Furthermore, the stability of the learning algorithms for control and identification purposes are proved by using appropriate Lyapunov functions. In addition to its well-known feature which is robustness, the most significant advantage of the proposed learning algorithm for the identification purposes is that the algorithm has a closed form, and thus it is easier to implement in real-time compared to other existing methods.

Fuzzy Neural Networks for Real Time Control Applications
http://dx.doi.org/10.1016/B978-0-12-802687-8.00007-4

Keywords

Sliding mode control theory based adaptation rules, Type-2 fuzzy neural networks, Feedback error learning controller, Identification, Control

7.1 INTRODUCTION

SMC is a robust control algorithm that can handle nonlinear dynamic systems in the presence of bounded input disturbances and uncertainties. However, this controller suffers from several drawbacks:

1. Its basic version is sensitive to noise.
2. Its corresponding control signal exhibits high frequency oscillations.
3. In the design of an SMC, the nominal values of the nonlinear functions that exist in the dynamic of the system must be known.

In order to overcome the problems mentioned above, intelligent methods such as ANNs, fuzzy logic, and FNNs can be preferred. With its proven general function approximation property, flexibility and capability of using expert knowledge of FNNs, this structure is one of the most important structures to overcome the drawbacks of SMC [1]. The FNN-based methods are known to result in less chattering in control signal and eliminate the knowledge about the nominal functions of the dynamic model of the system. The use of fuzzy systems in parallel with SMC approaches has been considered in number of papers. Sliding mode fuzzy controllers have the advantages of both fuzzy systems and SMCs. Fuzzy systems can be used as powerful and flexible approximators, and the sliding mode approach adds the possibility of thorough stability analysis and robustness against uncertainties both for the system and the adaptation laws [2–4].

In this chapter, not only is the fusion of T2FNNs with SMC theory-based learning algorithms in a control scheme considered, but also SMC theory-based learning algorithms are used to train the parameters of a T2FNN as an identifier. This combination makes it possible to train the parameters of T2FNNs with guaranteed stability analysis. In other words, during the training of the parameters of the T2FNN using SMC theory-based learning algorithms, it is guaranteed that the identification error will converge to zero, and the parameters of T2FNN will not diverge.

7.2 IDENTIFICATION DESIGN

A first-order interval type-2 TSK fuzzy *if-then* rule base with r input variables is used in this chapter for identification purposes. The consequent part of which is composed of crisp numbers, and its premise part is type-2

MFs. The definitions and the structure of this type of fuzzy system can be found in the previous chapters, but here we summarize it for convenience. The rth rule is as follows:

$$R_r: \text{ If } x_1 \text{ is } \tilde{A}_{1j} \ \ldots \text{ and } x_i \text{ is } \tilde{A}_{ik} \text{ and} \ldots \text{ and } x_I \text{ is } \tilde{A}_{Il} \text{ then}$$

$$f_r = \sum_{i=1}^{I} a_{ri} x_i + b_r \tag{7.1}$$

where $x_i (i = 1, \ldots, I)$ are the inputs of the type-2 TSK model, $\tilde{A}_{ik}$ is the kth type-2 fuzzy MF ($k = 1, \ldots, K$) corresponding to the input ith variable, and K is the number of MFs for the ith input that can be different for each input. The parameters a_r and b_r stand for the consequent part and $f_r (r = 1, \ldots, N)$ is the output function. The MFs used in this chapter are type-2 fuzzy Gaussian MFs and elliptic type-2 MFs. The upper and lower type-2 fuzzy Gaussian MFs with an uncertain standard deviation can be represented as follows:

$$\overline{\mu}_{ik}(x_i) = \exp\left(-\frac{1}{2}\frac{(x_i - c_{ik})^2}{\overline{\sigma}_{ik}^2}\right) \tag{7.2}$$

$$\underline{\mu}_{ik}(x_i) = \exp\left(-\frac{1}{2}\frac{(x_i - c_{ik})^2}{\underline{\sigma}_{ik}^2}\right) \tag{7.3}$$

where c_{ik} is the center value of the kth type-2 fuzzy set for the ith input. The parameters $\overline{\sigma}_{ik}$ and $\underline{\sigma}_{ik}$ are standard deviations for the upper and lower MFs, and the upper and lower type-2 fuzzy Elliptic MFs can be represented as follows:

$$\underline{\mu}_{ik} = \left(1 - \left|\frac{x_i - c_{ik}}{d_{ik}}\right|^{a_{2,ik}}\right)^{1/a_{2,ik}} H(x_i, c_{ik}, d_{ik}) \tag{7.4}$$

$$\overline{\mu}_{ik} = \left(1 - \left|\frac{x_i - c_{ik}}{d_{ik}}\right|^{a_{1,ik}}\right)^{1/a_{1,ik}} H(x_i, c_{ik}, d_{ik}) \tag{7.5}$$

in which $\underline{\mu}_{ik}$ and $\overline{\mu}_{ik}$ are the lower and upper bound of kth MF considered for the ith input and $H(x_i, c_{ik}, d_{ik})$ is the rectangular function defined as:

$$H(x_i, c, d) = \begin{cases} 1 & c - d < x_i < c + d \\ 0 & \text{otherwise} \end{cases}$$

The structure used in this investigation is called the A2-C0 fuzzy system [5]. In such a structure, first, the lower and upper membership degrees $\underline{\mu}$ and $\overline{\mu}$ are determined for each input signal being fed to the system. Next, the firing strengths of the rules using the *prod* t-norm operator are calculated as follows:

$$\underline{w}_r = \underline{\mu}_{\tilde{A}1}(x_1) * \underline{\mu}_{\tilde{A}2}(x_2) * \cdots \underline{\mu}_{\tilde{A}I}(x_I)$$
$$\overline{w}_r = \overline{\mu}_{\tilde{A}1}(x_1) * \overline{\mu}_{\tilde{A}2}(x_2) * \cdots \overline{\mu}_{\tilde{A}I}(x_I) \tag{7.6}$$

The consequent part corresponding to each fuzzy rule is a linear combination of the inputs $x_1, x_2, \ldots, x_I$. This linear function is called f_r and is defined as in (7.1). The output of the network is approximated as follows:

$$y_N = q\sum_{r=1}^{N} f_r\tilde{\underline{w}}_r + (1-q)\sum_{r=1}^{N} f_r\tilde{\overline{w}}_r \tag{7.7}$$

where $\tilde{\underline{w}}_r$ and $\tilde{\overline{w}}_r$ are the normalized values of the lower and upper output signals from the second hidden layer of the network as follows:

$$\tilde{\underline{w}}_r = \frac{\underline{w}_r}{\sum_{i=1}^{N}\underline{w}_r} \text{ and } \tilde{\overline{w}}_r = \frac{\overline{w}_r}{\sum_{i=1}^{N}\overline{w}_r} \tag{7.8}$$

The design parameter, q, is a coefficient which determines the sharing of the lower and the upper firing levels of each fired rule. This parameter can be a constant (equal to 0.5 in most cases) or a time-varying parameter. In this investigation, the latter is preferred. In other words, the parameter update rules and the proof of the stability of the learning process are given for the case of a time-varying q.

The following vectors can be specified:

$$\tilde{\underline{W}}(t) = \left[\tilde{\underline{w}}_1(t)\ \tilde{\underline{w}}_2(t) \ldots \tilde{\underline{w}}_N(t)\right]^{\mathrm{T}},$$
$$\tilde{\overline{W}}(t) = \left[\tilde{\overline{w}}_1(t)\ \tilde{\overline{w}}_2(t) \ldots \tilde{\overline{w}}_N(t)\right]^{\mathrm{T}} \text{ and } F = [f_1\, f_2\, \ldots f_N]$$

The following assumptions have been used in this investigation: The time derivative of both the input and output signals can be considered bounded:

$$|\dot{x}_i(t)| \leq B_{\dot{x}}, \quad \min(x_i^2(t)) = B_{x^2}, \ (i = 1 \ldots I) \text{ and } |\dot{y}(t)| \leq B_{\dot{y}} \quad \forall t \tag{7.9}$$

where $B_{\dot{x}}$, B_{x^2}, and $B_{\dot{y}}$ are assumed to be some known positive constants.

7.2.1 Identification Using Gaussian Type-2 MF with Uncertain σ

Since Gaussian type-2 MF with uncertain σ is the only MF that does not have any undifferentiable point, it is one of the first choices in type-2 fuzzy control and identification applications. The error between the measured output of the system and the output of the T2FNN can be defined as a time-varying sliding surface. In this section, it is shown that under certain conditions, it is guaranteed that the system will keep on being on the sliding surface. The use of *sign* function results in finite time convergence of the sliding surface to zero, which, in turn forces the output of the network, $y_N(t)$, to perfectly follow the desired output signal, $y(t)$, for all time $t > t_h$. The time instant t_h is defined to be the hitting time for being $e(t) = 0$. The sliding surface is defined as follows:

$$S(e(t)) = e(t) = y_N(t) - y(t) = 0 \tag{7.10}$$

Definition: A sliding motion will appear on the sliding manifold $S(e(t)) = e(t) = 0$ after a finite time t_h, if the condition $S(t)\dot{S}(t) < 0$ is satisfied for all t in some nontrivial semi-open subinterval of time of the form $[t, t_h) \subset (0, t_h)$.

The main goal of the design of the adaptation laws is to design an online learning algorithm for the parameters of the T2FNN, such that the sliding mode condition of the above definition is enforced.

7.2.1.1 Parameter Update Rules for the T2FNN

The parameter update rules for the T2FNN are given by the following theorem.

Theorem 7.1. *If the adaptation laws for the parameters of the considered T2FNN are chosen as [6]:*

$$\dot{c}_{ik} = \dot{x}_i + (x_i - c_{ik})\alpha_1 \operatorname{sgn}(e) \tag{7.11}$$

$$\dot{\underline{\sigma}}_{ik} = -\left(\underline{\sigma}_{ik} + \frac{\underline{\sigma}_{ik}^3}{(x_i - c_{ik})^2}\right)\alpha_1 \operatorname{sgn}(e) \tag{7.12}$$

$$\dot{\overline{\sigma}}_{ik} = -\left(\overline{\sigma}_{ik} + \frac{\overline{\sigma}_{ik}^3}{(x_i - c_{ik})^2}\right)\alpha_1 \operatorname{sgn}(e) \tag{7.13}$$

$$\dot{a}_{ri} = -x_i \frac{q\underline{\tilde{w}}_r + (1-q)\bar{\tilde{w}}_r}{\left(q\underline{\tilde{w}}_r + (1-q)\bar{\tilde{w}}_r\right)^T \left(q\underline{\tilde{w}}_r + (1-q)\bar{\tilde{w}}_r\right)} \alpha \operatorname{sgn}(e) \tag{7.14}$$

$$\dot{b}_r = -\frac{q\underline{\tilde{w}}_r + (1-q)\bar{\tilde{w}}_r}{\left(q\underline{\tilde{w}}_r + (1-q)\bar{\tilde{w}}_r\right)^T \left(q\underline{\tilde{w}}_r + (1-q)\bar{\tilde{w}}_r\right)} \alpha \operatorname{sgn}(e) \tag{7.15}$$

$$\dot{q} = -\frac{1}{F(\underline{\widetilde{W}} - \overline{\widetilde{W}})^{\mathrm{T}}} \alpha \operatorname{sgn}(e) \tag{7.16}$$

where α is an adaptive learning rate with an adaptation law as follows:

$$\dot{\alpha} = \gamma(I+2) \mid e \mid - \nu\gamma\alpha, \quad 0 < \gamma, \nu \tag{7.17}$$

then, given an arbitrary initial condition $e(0)$, the learning error $e(t)$ will converge to zero within a finite time t_h.

Remark 7.1. The adaptation laws of (7.17) show that the learning rate does not have a fixed value and its value is evolving during identification. The existence of this adaptation law makes it possible to choose a small initial value for α and it grows based on the requirement of identification during the training phase. Note that in (7.17), the parameter γ has a small positive real value that is interpreted as the learning rate for the adaptive learning rate. As can be seen for the adaptation law, the first term of the adaptation law of (7.17) is always positive, which may cause bursting in the parameter α. In order to avoid this phenomenon, the second term is added to the adaptation law, which avoids a possible parameter bursting in α. The value of ν should be selected very small to keep it from interrupting the adaptation mechanism.

Remark 7.2. Another benefit of the adaptive learning rate using (7.17) is that the upper bound of the states of the system does not need to be known as a priori and the knowledge of their existence is enough.

Remark 7.3. It is possible that the denumerator in the adaptation laws of (7.11)-(7.16) become zero, which may cause instability in the system. In order to avoid this, the denominator should be equal to a small number (e.g. 0.001) when its calculated value is smaller than this threshold.

In order to avoid the possibility of high-frequency oscillations in the control input, which is called *chattering*, the following are the two common methods used [7]:

1. Using a saturation function instead of the signum function,
2. Using a boundary layer so that an equivalent control replaces the corrective one when the system is inside this layer.

In order to reduce the chattering effect, the following function is used a small number, e.g. $\delta_s = 0.05$ instead of the signum function:

$$\operatorname{sgn}(e) := \frac{e}{|e| + \delta_s} \tag{7.18}$$

Remark 7.4. In different tests it is observed that the output of the T2FNN is quite sensitive to the changes in the consequent part parameters. Hence, it is logical to use larger values for the parameters of the consequent part and smaller values for the parameters of the antecedent part. Thus, a smaller value (α_1) is chosen for the antecedent parts.

7.2.1.2 Proof of Theorem 7.1

The time derivative of (7.8) is calculated as follows:

$$\dot{\underline{\tilde{w}}}_r = -\underline{\tilde{w}}_r\underline{K}_r + \underline{\tilde{w}}_r\sum_{r=1}^{N}\underline{\tilde{w}}_r\underline{K}_r;\quad \dot{\overline{\tilde{w}}}_r = -\overline{\tilde{w}}_r\overline{K}_r + \overline{\tilde{w}}_r\sum_{r=1}^{N}\overline{\tilde{w}}_r\overline{K}_r \tag{7.19}$$

where

$$\underline{A}_{ik} = \frac{x_i - c_{ik}}{\underline{\sigma}_{ik}} \text{ and } \overline{A}_{ik} = \frac{x_i - c_{ik}}{\overline{\sigma}_{ik}}$$

$$\underline{K}_r = \sum_{i=1}^{I}\underline{A}_{ik}\dot{\underline{A}}_{ik} \text{ and } \overline{K}_r = \sum_{i=1}^{I}\overline{A}_{ik}\dot{\overline{A}}_{ik}$$

Applying (7.11)-(7.13) into the equations above, the following equation is obtained:

$$\overline{K}_r = \underline{K}_r = \sum_{i=1}^{I}\overline{A}_{ik}\dot{\overline{A}}_{ik} = \sum_{i=1}^{I}\underline{A}_{ik}\dot{\underline{A}}_{ik} = I\alpha\operatorname{sgn}(e) \tag{7.20}$$

The following Lyapunov function is used to prove the stability of the proposed adaptation laws:

$$V = \frac{1}{2}e^2 + \frac{1}{2\gamma}(\alpha - \alpha^*)^2,\quad 0 < \gamma \tag{7.21}$$

The stability conditions require the time derivative of the Lyapunov function to be negative. The time derivative of (7.21) can be calculated as follows:

$$\dot{V} = \dot{e}e + \frac{1}{\gamma}\dot{\alpha}(\alpha - \alpha^*) = e(\dot{y}_N - \dot{y}) + \frac{1}{\gamma}\dot{\alpha}(\alpha - \alpha^*) \tag{7.22}$$

The time derivative of the Lyapunov function requires the time derivative of y_N to be calculated. Hence, the time derivative of (7.7) is obtained as follows:

$$\begin{aligned}\dot{y}_N = \dot{q}\sum_{r=1}^{N} f_r\tilde{\underline{w}}_r + q\sum_{r=1}^{N}(\dot{f}_r\tilde{\underline{w}}_r + f_r\dot{\tilde{\underline{w}}}_r) - \dot{q}\sum_{r=1}^{N} f_r\tilde{\overline{w}}_r \\ + (1-q)\sum_{r=1}^{N}(\dot{f}_r\tilde{\overline{w}}_r + f_r\dot{\tilde{\overline{w}}}_r)\end{aligned} \tag{7.23}$$

By using (7.19), (7.20), and (7.23), the following term can be obtained:

$$\begin{aligned}\dot{y}_N =& \dot{q}\sum_{r=1}^{N} f_r\tilde{\underline{w}}_r + q\sum_{r=1}^{N}\left(\dot{f}_r\tilde{\underline{w}}_r + f_r\left(-\tilde{\underline{w}}_r\underline{K}_r + \tilde{\underline{w}}_r\sum_{r=1}^{N}\tilde{\underline{w}}_r\underline{K}_r\right)\right) \\ &- \dot{q}\sum_{r=1}^{N} f_r\tilde{\overline{w}}_r + (1-q)\sum_{r=1}^{N}\left(\dot{f}_r\tilde{\overline{w}}_r + f_r\left(-\tilde{\overline{w}}_r\overline{K}_r + \tilde{\overline{w}}_r\sum_{r=1}^{N}\tilde{\overline{w}}_r\overline{K}_r\right)\right) \\ =& \dot{q}\sum_{r=1}^{N} f_r\tilde{\underline{w}}_r + q\sum_{r=1}^{N}\left(\dot{f}_r\tilde{\underline{w}}_r - I\alpha\,\mathrm{sgn}(e) f_r\left(\tilde{\underline{w}}_r - \tilde{\underline{w}}_r\sum_{r=1}^{N}\tilde{\underline{w}}_r\right)\right) \\ &- \dot{q}\sum_{r=1}^{N} f_r\tilde{\overline{w}}_r + (1-q)\sum_{r=1}^{N}\left(\dot{f}_r\tilde{\overline{w}}_r - I\alpha\,\mathrm{sgn}(e) f_r\left(\tilde{\overline{w}}_r - \tilde{\overline{w}}_r\sum_{r=1}^{N}\tilde{\overline{w}}_r\right)\right)\end{aligned}$$

Equation (7.24) is correct by definition:

$$\sum_{r=1}^{N}\tilde{\underline{w}}_r = 1 \text{ and } \sum_{r=1}^{N}\tilde{\overline{w}}_r = 1 \tag{7.24}$$

It is possible to use (7.14), (7.15), (7.16), and (7.24) to obtain the following equation:

$$\dot{y}_N = -\frac{1}{F(\underline{\tilde{W}} - \tilde{\overline{W}})^T}\alpha \mathrm{sgn}\,(e)\sum_{r=1}^{N} f_r(\underline{\tilde{w}}_r - \tilde{\overline{w}}_r) + \sum_{r=1}^{N}\dot{f}_r\Big(q\underline{\tilde{w}}_r + (1-q)\tilde{\overline{w}}_r\Big)$$
$$= -\alpha \mathrm{sgn}\,(e) + \sum_{r=1}^{N}\Bigg[\Big(\sum_{i=1}^{I}(\dot{a}_{ri}x_i + a_{ri}\dot{x}_i) + \dot{b}_r\Big)\Big(q\underline{\tilde{w}}_r + (1-q)\tilde{\overline{w}}_r\Big)\Bigg] \tag{7.25}$$

By substituting (7.25) into (7.22), the following equation can be obtained:

$$\dot{V} = \dot{e}e + \frac{1}{\gamma}\dot{\alpha}(\alpha - \alpha^*) = e(\dot{y}_N - \dot{y}) + \frac{1}{\gamma}\dot{\alpha}(\alpha - \alpha^*)$$
$$= e\Bigg[-\alpha \mathrm{sgn}\,(e) + \sum_{r=1}^{N}\Bigg[\Big(\sum_{i=1}^{I}(\dot{a}_{ri}x_i + a_{ri}\dot{x}_i) + \dot{b}_r\Big)\Big(q\underline{\tilde{w}}_r + (1-q)\tilde{\overline{w}}_r\Big)\Bigg] - \dot{y}\Bigg] + \frac{1}{\gamma}\dot{\alpha}(\alpha - \alpha^*) \tag{7.26}$$

By inserting the adaptation laws of (7.14)-(7.16) into (7.26), the following equation is obtained:

$$\dot{V} = e\Bigg[-\alpha \mathrm{sgn}\,(e) + \sum_{r=1}^{N}\Bigg[\Big(\sum_{i=1}^{I}\Big(-\big(x_i\alpha \mathrm{sgn}\,(e)\frac{(q\underline{\tilde{w}}_r + (1-q)\tilde{\overline{w}}_r)}{(q\underline{\tilde{w}}_r + (1-q)\tilde{\overline{w}}_r)^{\mathrm{T}}(q\underline{\tilde{w}}_r + (1-q)\tilde{\overline{w}}_r)}\big)x_i + a_{ri}\dot{x}_i\Big) - \alpha \mathrm{sgn}\,(e)\frac{(q\underline{\tilde{w}}_r + (1-q)\tilde{\overline{w}}_r)}{(q\underline{\tilde{w}}_r + (1-q)\tilde{\overline{w}}_r)^{\mathrm{T}}(q\underline{\tilde{w}}_r + (1-q)\tilde{\overline{w}}_r)}\Big)(q\underline{\tilde{w}}_r + (1-q)\tilde{\overline{w}}_r)\Bigg] - \dot{y}\Bigg] + \frac{1}{\gamma}\dot{\alpha}(\alpha - \alpha^*) \tag{7.27}$$

$$= e\Bigg[- \alpha \mathrm{sgn}\,(e) + \sum_{r=1}^{N}\Bigg[\sum_{i=1}^{I}\Big(- \alpha \mathrm{sgn}\,(e)\, x_i^2 + a_{ri}\dot{x}_i\big(q\underline{\tilde{w}}_r + (1-q)\tilde{\overline{w}}_r\big)\Big) - \alpha \mathrm{sgn}\,(e)\Bigg] - \dot{y}\Bigg] + \frac{1}{\gamma}\dot{\alpha}(\alpha - \alpha^*) \tag{7.28}$$

Furthermore:

$$\dot{V} = -\mid e \mid 2\alpha + e\Bigg[\sum_{r=1}^{N}\Bigg[\sum_{i=1}^{I}\Big(- \alpha \mathrm{sgn}\,(e)\, x_i^2 + a_{ri}\dot{x}_i\big(q\underline{\tilde{w}}_r + (1-q)\tilde{\overline{w}}_r\big)\Big)\Bigg] - \dot{y}\Bigg] + \frac{1}{\gamma}\dot{\alpha}(\alpha - \alpha^*) \tag{7.29}$$

Using the upper bounds of the signals as:

$$\|x_i\| \leq B_{x^2}, \quad \|\dot{y}\| \leq B_{\dot{y}}, \quad \|\dot{x}_i\| \leq B_{\dot{x}}, \quad \|a_r\| \leq B_a \tag{7.30}$$

we have the following inequalities:

$$\dot{V} < -\mid e \mid 2\alpha + \mid e \mid\Bigg[\sum_{r=1}^{N}\Bigg[\sum_{i=1}^{I}\big(- \alpha B_{x^2} + B_a B_{\dot{x}}(q\underline{\tilde{w}}_r + (1-q)\tilde{\overline{w}}_r)\big)\Bigg] + B_{\dot{y}}\Bigg] + \frac{1}{\gamma}\dot{\alpha}(\alpha - \alpha^*)$$

$$< -\mid e \mid 2\alpha + \mid e \mid\Bigg[\sum_{r=1}^{N}\Big[- I\alpha B_{x^2} + IB_a B_{\dot{x}}(q\underline{\tilde{w}}_r + (1-q)\tilde{\overline{w}}_r)\Big] + B_{\dot{y}}\Bigg] + \frac{1}{\gamma}\dot{\alpha}(\alpha - \alpha^*)$$

$$< -\mid e \mid 2\alpha + \mid e \mid\Bigg[- I\alpha B_{x^2} + IB_a B_{\dot{x}}(q\sum_{r=1}^{N}\underline{\tilde{w}}_r$$

$$+ (1-q)\sum_{r=1}^{N}\tilde{\overline{w}}_r) + B_{\dot{y}}\Bigg] + \frac{1}{\gamma}\dot{\alpha}(\alpha-\alpha^*)$$

$$< -\mid e\mid \left(2\alpha + I\alpha B_{x^2}\right) + \mid e\mid \left[IB_a B_{\dot{x}} + B_{\dot{y}}\right]$$

$$+\frac{1}{\gamma}\dot{\alpha}(\alpha-\alpha^*)$$

The parameter α^* is considered to be an unknown parameter that is determined during the adaptation of the learning rate and its true value is considered to satisfy the following inequality:

$$\alpha^* \geq \frac{2(IB_a B_{\dot{x}} + B_{\dot{y}})}{2 + IB_{x^2}}$$

Hence, using an adaptation law for the learning rate makes it possible to estimate the parameters of the T2FNN with less a priori knowledge. The adaptation law of the learning rate is obtained in the next few steps:

$$\begin{aligned}\dot{V} \leq & -\mid e\mid (2\alpha + I\alpha B_{x^2}) + \mid e\mid (IB_a B_{\dot{x}} + B_{\dot{y}}) \\ & + (2 + IB_{x^2})\alpha^* \mid e\mid - (2+IB_{x^2})\alpha^* \mid e\mid \\ & + \frac{1}{\gamma}(\alpha-\alpha^*)\dot{\alpha}\end{aligned} \tag{7.31}$$

$$\begin{aligned}\dot{V} \leq & \mid e\mid (IB_a B_{\dot{x}} + B_{\dot{y}}) - (2 + IB_{x^2})\alpha^* \mid e\mid \\ & - (2+IB_{x^2})(\alpha^*-\alpha)\mid e\mid + \frac{1}{\gamma}(\alpha-\alpha^*)\dot{\alpha}\end{aligned} \tag{7.32}$$

and further:

$$\begin{aligned}\dot{V} \leq & \mid e\mid (IB_a B_{\dot{x}} + B_{\dot{y}}) - (2 + IB_{x^2})\alpha^* \mid e\mid \\ & + (\alpha^*-\alpha)\left[(2+IB_{x^2})\mid e\mid - \frac{1}{\gamma}\dot{\alpha}\right]\end{aligned} \tag{7.33}$$

Using the adaptation law for the adaptive learning rate (α) as:

$$\dot{\alpha} = (2+IB_{x^2})\gamma \mid e\mid - \nu\gamma\alpha \tag{7.34}$$

in which ν has a small real value, the time derivative of the Lyapunov function can be rewritten as:

$$\begin{aligned}\dot{V} \leq & \mid e \mid (IB_a B_{\dot{x}} + B_{\dot{y}}) \\ & - (2 + IB_{x^2})\alpha^* \mid e \mid + (\alpha^* - \alpha)\nu\alpha \end{aligned} \tag{7.35}$$

so that:

$$\begin{aligned}\dot{V} \leq & \mid e \mid (IB_a B_{\dot{x}} + B_{\dot{y}}) \\ & - (2 + IB_{x^2})\alpha^* \mid e \mid - \nu(\alpha - \frac{\alpha^*}{2})^2 + \frac{\nu\alpha^{*2}}{4} \end{aligned} \tag{7.36}$$

considering the fact that $\alpha^* \geq \frac{2(IB_a B_{\dot{x}} + B_{\dot{y}})}{2 + IB_{x^2}}$, we have $\mid e \mid (IB_a B_{\dot{x}} + B_{\dot{y}}) - \frac{\alpha^*}{2}(2 + IB_{x^2}) \mid e \mid \leq 0$ and consequently:

$$\dot{V} \leq -\frac{\alpha^*}{2}(2 + IB_{x^2}) \mid e \mid + \frac{\nu\alpha^{*2}}{4} \tag{7.37}$$

Therefore, error converges to a very small region around zero in which $\mid e \mid \leq \frac{\alpha^* \nu}{2(2 + IB_{x^2})}$, and it remains there. It should also be noted that ν is a small user-defined positive number that can be selected as small as desired to make this neighborhood as narrow as requested by the user.

7.2.2 Identification Using T2FNN with Elliptic Type-2 MF

In this section, the SMC theory-based parameter update rules for the T2FNN with elliptic MFs are discussed. The introduced novel training algorithm is simple and computationally less expensive than the gradient-based methods, and the adaptation laws have closed forms. The stability of the proposed method is proved using the Lyapunov approach. Similar to the previous section, the error between the measured output of the system and the output of the T2FNN is defined as a time-varying sliding surface. Similar to identification using Gaussian type-2 MF, the sliding surface is defined as follows:

$$S(e(t)) = e(t) = y_N(t) - y(t) = 0 \tag{7.38}$$

The sliding surface for the nonlinear system is defined as:

$$S_p = e \tag{7.39}$$

7.2.2.1 Parameter Update Rules for the T2FNN

The parameter update rules for the T2FNN are given by the following theorem.

Theorem 7.2. *Consider a T2FNN structure that benefits from elliptic type-2 MF. Using the following assumptions:*

$$|x_i| < B_x, \quad x_i^2 < B_{x^2}, \quad |\dot{x}_i| < B_{\dot{x}}, \quad |\dot{y}| < B_{\dot{y}}$$

if the parameter update rules for the T2FNN are chosen as (7.40)-(7.47) *with the selection of the adaptive learning rate* (*α*) *as in* (7.47), *e*(*t*), *the error between the desired output and the output of the T2FNN converges asymptotically to zero in finite time and sliding motion will be achieved.*

$$\dot{a}_{2,ik} = \gamma_1 \left\{ \frac{ln\left(1 - \left|\frac{x_i - c_{ik}}{d_{ik}}\right|^{a_{2,ik}}\right)}{a_{2,ik}^2} + \frac{\left|\frac{x_i - c_{ik}}{d_{ik}}\right|^{a_{2,ik}} ln\left|\frac{x_i - c_{ik}}{d_{ik}}\right|}{a_{2,ik}(1 - T_{2,ik})} \right\}^{-1} H(x_i, c_{ik}, d_{ik}) \tag{7.40}$$

$$\dot{a}_{1,ik} = \gamma_1 \left\{ \frac{ln\left(1 - \left|\frac{x_i - c_{ik}}{d_{ik}}\right|^{a_{1,ik}}\right)}{a_{1,ik}^2} + \frac{\left|\frac{x_i - c_{ik}}{d_{ik}}\right|^{a_{1,ik}} ln\left|\frac{x_i - c_{ik}}{d_{ik}}\right|}{a_{1,ik}(1 - T_{1,ik})} \right\}^{-1} H(x_i, c_{ik}, d_{ik}) \tag{7.41}$$

$$\dot{c}_{ik} = -\gamma_1 \frac{|d_{ik}|^{a_{2,ik}}(1 - T_{2,ik})\mathrm{sgn}(x_i - c_{ik})}{|x_i - c_{ik}|^{a_{2,ik}-1}} \tag{7.42}$$

$$\dot{d}_{ik} = \gamma_1 \frac{(1 - T_{2,ik})|d_{ik}|^{a_{2,ik}+1}}{|x_i - c_{ik}|^{a_{2,ik}}} \mathrm{sgn}(d_{ik}) \tag{7.43}$$

$$\dot{b}_r = -\frac{q\underline{\widetilde{W}}_r + (1-q)\overline{\widetilde{W}}_r}{\Pi_r^{\mathrm{T}}\Pi_r} \alpha \mathrm{sign}(e) \tag{7.44}$$

in which:

$$\Pi_r = \left(\sum_{k=1}^{R}(q\underline{\widetilde{W}}_r + (1-q)\overline{\widetilde{W}}_r)\right) \tag{7.45}$$

$$\dot{a}_{ri} = -x_i \frac{q\underline{\widetilde{W}}_r + (1-q)\overline{\widetilde{W}}_r}{\Pi_r^{\mathrm{T}}\Pi_r}\alpha \mathrm{sign}(e) \tag{7.46}$$

$$\dot{\alpha} = \gamma(I+2) \mid e \mid -\nu\gamma\alpha, \;\; 0 < \gamma, \nu \tag{7.47}$$

In the above γ_1 and γ are the learning rates considered for the adaptation of the learning rate and they need to be selected as positive.

7.2.2.2 Proof of Theorem 7.2

As mentioned earlier, the elliptic type-2 MFs are used so that:

$$\underline{\mu}_{ik} = \left(1 - \left|\frac{x_i - c_{ik}}{d_{ik}}\right|^{a_{2,ik}}\right)^{1/a_{2,ik}} H(x_i, c_{ik}, d_{ik}) \tag{7.48}$$

in which $\underline{\mu}_{ik}$ is the lower bound of the kth MF considered for the ith input and $H(x_i, c_{ik}, d_{ik})$ is the rectangular function defined as:

$$H(x_i, c, d) = \begin{cases} 1 & c - d < x_i < c + d \\ 0 & \text{otherwise} \end{cases}$$

If $c_{ik} - d_{ik} < x_i < c_{ik} + d_{ik}$, then the following equation is valid:

$$ln(\underline{\mu}_{ik}) = \frac{1}{a_{2,ik}} ln\left(1 - \left|\frac{x_i - c_{ik}}{d_{ik}}\right|^{a_{2,ik}}\right) \tag{7.49}$$

Define $T_{2,ik} = \left|\frac{x_i - c_{ik}}{d_{ik}}\right|^{a_{2,ik}}$ so that:

$$\begin{aligned}\dot{T}_{2,ik} = {} & \dot{a}_{2,ik}\left|\frac{x_i - c_{ik}}{d_{ik}}\right|^{a_{2,ik}} ln\left|\frac{x_i - c_{ik}}{d_{ik}}\right| \\ & + a_{2,ik}\left(\frac{\dot{x}_i - \dot{c}_{ik}}{|d_{ik}|}\mathrm{sgn}(x_i - c_{ik})\right.\end{aligned}$$

$$- \dot{d}_{ik}\frac{|x_i - c_{ik}|}{d_{ik}^2}\text{sgn}(d_{ik})\Bigg)\left|\frac{x_i - c_{ik}}{d_{ik}}\right|^{a_{2,ik}-1} \tag{7.50}$$

and the time derivative of $\underline{\mu}_{ik}$ is achieved as:

$$\dot{\underline{\mu}}_{ik} = \frac{-\underline{\mu}_{ik}\dot{a}_{2,ik}}{a_{2,ik}^2} ln\left(1 - \left|\frac{x_i - c_{ik}}{d_{ik}}\right|^{a_{2,ik}}\right) - \frac{\underline{\mu}_{ik}}{a_{2,ik}}\frac{\dot{T}_{2,ik}}{1 - T_{2,ik}} \tag{7.51}$$

The upper bound of the kth type-2 MFs for the ith input is defined as:

$$\overline{\mu}_{ik} = \left(1 - \left|\frac{x_i - c_{ik}}{d_{ik}}\right|^{a_{1,ik}}\right)^{1/a_{1,ik}} H(x_i, c_{ik}, d_{ik}) \tag{7.52}$$

Define $T_{1,ik} = \left|\frac{x_i - c_{ik}}{d_{ik}}\right|^{a_{1,ik}}$ so that:

$$\begin{aligned}\dot{T}_{1,ik} = {} & \dot{a}_{1,ik}\left|\frac{x_i - c_{ik}}{d_{ik}}\right|^{a_{1,ik}} ln\left|\frac{x_i - c_{ik}}{d_{ik}}\right| \\ & + a_{1,ik}\Bigg(\frac{\dot{x}_i - \dot{c}_{ik}}{|d_{ik}|}\text{sgn}(x_i - c_{ik}) \\ & - \dot{d}_{ik}\frac{|x_i - c_{ik}|}{d_{ik}^2}\text{sgn}(d_{ik})\Bigg)\left|\frac{x_i - c_{ik}}{d_{ik}}\right|^{a_{1,ik}-1}\end{aligned} \tag{7.53}$$

and further:

$$\dot{\overline{\mu}}_{ik} = \frac{-\overline{\mu}_{ik}\dot{a}_{1,ik}}{a_{1,ik}^2} ln\left(1 - \left|\frac{x_1 - c_{ik}}{d_{ik}}\right|^{a_{1,ik}}\right) - \frac{\overline{\mu}_{ik}}{a_{1,ik}}\frac{\dot{T}_{1,ik}}{1 - T_{1,ik}} \tag{7.54}$$

$$\begin{aligned}\frac{\overline{\mu}_{ik}}{1 - T_{1,ik}}\left|\frac{x_i - c_{ik}}{d_{ik}}\right|^{a_{1,ik}-1} & = \frac{\overline{\mu}_{ik} - \underline{\mu}_{ik} + \underline{\mu}_{ik}}{1 - T_{2,ik} + T_{2,ik} - T_{1,ik}}\left(\left|\frac{x_i - c_{ik}}{d_{ik}}\right|^{a_{2,ik}-1}\right. \\ & \quad \left. + \left|\frac{x_i - c_{ik}}{d_{ik}}\right|^{a_{1,ik}-1} - \left|\frac{x_i - c_{ik}}{d_{ik}}\right|^{a_{2,ik}-1}\right) \\ & = \frac{\underline{\mu}_{ik}}{(1 - T_{2,ik})}\left|\frac{x_i - c_{ik}}{d_{ik}}\right|^{a_{2,ik}-1}\left(1 + \frac{\overline{\mu}_{ik} - \underline{\mu}_{ik}}{\underline{\mu}_{ik}}\right) \times\end{aligned}$$

$$\left(1+\left(\left|\frac{x_i-c_{ik}}{d_{ik}}\right|^{a_{1,ik}-1}-\left|\frac{x_1-c_{ik}}{d_{ik}}\right|^{a_{2,ik}-1}\right)\times\right.$$
$$\left.\left(\left|\frac{x_i-c_{ik}}{d_{ik}}\right|^{a_{2,ik}-1}\right)^{-1}\right)\left(\frac{1}{1-\frac{T_{1,ik}-T_{2,ik}}{1-T_{2,ik}}}\right) \tag{7.55}$$

Considering the fact that the MFs have nonzero value only if $c_{ik}-d_{ik}<x_i<c_{ik}+d_{ik}$ we have:

$$\begin{aligned} 0 &< T_{1,ik} < 1 \\ T_{2,ik} &< T_{1,ik} \end{aligned} \tag{7.56}$$

so that:

$$0<T_{1,ik}-T_{2,ik}<1-T_{2,ik} \Rightarrow 0<\frac{T_{1,ik}-T_{2,ik}}{1-T_{2,ik}}<1 \tag{7.57}$$

This means it is possible to use the Maclaurin series as:

$$\frac{\overline{\mu}_{ik}}{1-T_{1,ik}}\left|\frac{x_i-c_{ik}}{d_{ik}}\right|^{a_{1,ik}-1}=\frac{\underline{\mu}_{ik}}{(1-T_{2,ik})}\left|\frac{x_i-c_{ik}}{d_{ik}}\right|^{a_{2,ik}-1}\left(1+\frac{\overline{\mu}_{ik}-\underline{\mu}_{ik}}{\underline{\mu}_{ik}}\right)\times$$
$$\left(1+\frac{T_{1,ik}-T_{2,ik}}{1-T_{2,ik}}+\left(\frac{T_{1,ik}-T_{2,ik}}{1-T_{2,ik}}\right)^2+H.O.T.\right)\times$$
$$\left(1+\left(\left|\frac{x_i-c_{ik}}{d_{ik}}\right|^{a_{1,ik}-1}-\left|\frac{x_i-c_{ik}}{d_{ik}}\right|^{a_{2,ik}-1}\right)\times\left(\left|\frac{x_i-c_{ik}}{d_{ik}}\right|^{a_{2,ik}-1}\right)^{-1}\right) \tag{7.58}$$

It should also be noted that:

$$\begin{aligned} \frac{\overline{\mu}_{ik}}{(1-T_{1,ik})}\left|\frac{x_i-c_{ik}}{d_{ik}}\right|^{a_{1,ik}-1} &= \frac{\underline{\mu}_{ik}}{(1-T_{2,ik})}\left|\frac{x_i-c_{ik}}{d_{ik}}\right|^{a_{2,ik}-1}(1+\Delta_{ik}) \\ &= \frac{\underline{\mu}_{ik}}{(1-T_{2,ik})}\left|\frac{x_i-c_{ik}}{d_{ik}}\right|^{a_{2,ik}-1}+\Delta_{ik}^{'} \end{aligned} \tag{7.59}$$

in which:

$$\Delta_{ik}^{'} = \frac{\underline{\mu}_{ik}}{(1 - T_{2,ik})} \left| \frac{x_i - c_{ik}}{d_{ik}} \right|^{a_{2,ik}-1} \Delta_{ik} \tag{7.60}$$

and:

$$\begin{aligned} \Delta_{ik} = & \left(1 + \frac{\overline{\mu}_{ik} - \underline{\mu}_{ik}}{\underline{\mu}_{ik}}\right) \times \\ & \left(1 + \frac{T_{1,ik} - T_{2,ik}}{1 - T_{2,ik}} + \left(\frac{T_{1,ik} - T_{2,ik}}{1 - T_{2,ik}}\right)^2 + \text{H.O.T.}\right) \times \\ & \left(1 + \left(\left|\frac{x_2 - c_{ik}}{d_{ik}}\right|^{a_{1,ik}-1} - \left|\frac{x_i - c_{ik}}{d_{ik}}\right|^{a_{2,ik}-1}\right) \times \right. \\ & \left. \left(\left|\frac{x_i - c_{ik}}{d_{ik}}\right|^{a_{2,ik}-1}\right)^{-1}\right) - 1 \end{aligned} \tag{7.61}$$

Considering the adaptation laws as in (7.40)-(7.47), we have:

$$\dot{\underline{\mu}}_{ik} = -\gamma \underline{\mu}_{ik} - \frac{\underline{\mu}_{ik}}{1 - T_{2,ik}} \left| \frac{x_i - c_{ik}}{d_{ik}} \right| \dot{x}_i \text{sgn}(x_i - c_{ik}) \tag{7.62}$$

$$\dot{\overline{\mu}}_{ik} = -\gamma \overline{\mu}_{ik} + \Lambda_{ik} \tag{7.63}$$

where Λ_{ik} is defined as:

$$\Lambda_{ik} = \gamma \overline{\mu}_{ik} - \gamma \underline{\mu}_{ik} - \frac{\overline{\mu}_{ik}}{1 - T_{1,ik}} \left| \frac{x_i - c_{ik}}{d_{ik}} \right| \dot{x}_i \text{sgn}(x_i - c_{ik}) \tag{7.64}$$

and further:

$$\dot{\underline{w}}_r = -\gamma \underline{w}_r \text{sgn}(e) + \Delta \underline{w}_r \tag{7.65}$$

and

$$\dot{\overline{w}}_r = -\gamma \overline{w}_r \text{sgn}(e) + \Delta \overline{w}_r \tag{7.66}$$

where $\Delta \underline{w}_r$ and $\Delta \overline{w}_r$ are bounded and satisfy the following equation:

$$|\Delta \underline{w}_r|,\ |\Delta \overline{w}_r| < B_{\Delta w} \tag{7.67}$$

By using the following Lyapunov function, the stability condition is analyzed:

$$V = \frac{1}{2}e^2 + \frac{1}{2\gamma}(\alpha - \alpha^*)^2, \quad 0 < \gamma \tag{7.68}$$

The time derivative of (7.68) can be calculated as follows:

$$\dot{V} = \dot{e}e + \frac{1}{\gamma}\dot{\alpha}(\alpha - \alpha^*) = e(\dot{y}_N - \dot{y}) + \frac{1}{\gamma}\dot{\alpha}(\alpha - \alpha^*) \tag{7.69}$$

By taking the time-derivative of (7.7), the following term is obtained:

$$\begin{aligned}\dot{y}_N = {} & \dot{q}\sum_{r=1}^{N} f_r\tilde{\underline{w}}_r + q\sum_{r=1}^{N}(\dot{f}_r\tilde{\underline{w}}_r + f_r\dot{\tilde{\underline{w}}}_r) - \dot{q}\sum_{r=1}^{N} f_r\tilde{\overline{w}}_r \\ & + (1-q)\sum_{r=1}^{N}(\dot{f}_r\tilde{\overline{w}}_r + f_r\dot{\tilde{\overline{w}}}_r)\end{aligned} \tag{7.70}$$

Using the adaptation laws as in (7.40)-(7.47), the following term can be obtained:

$$\begin{aligned}\dot{y}_N = {} & \dot{q}\sum_{r=1}^{N} f_r\tilde{\underline{w}}_r + q\sum_{r=1}^{N}(\dot{f}_r\tilde{\underline{w}}_r + f_r\Delta\tilde{\underline{w}}_r) - \dot{q}\sum_{r=1}^{N} f_r\tilde{\overline{w}}_r \\ & + (1-q)\sum_{r=1}^{N}(\dot{f}_r\tilde{\overline{w}}_r + f_r\Delta\tilde{\overline{w}}_r)\end{aligned} \tag{7.71}$$

and further we have:

$$\begin{aligned}\dot{y}_N = {} & -\frac{1}{F(\underline{\tilde{W}} - \overline{\tilde{W}})^T}\alpha \operatorname{sgn}(e)\sum_{r=1}^{N} f_r(\tilde{\underline{w}}_r - \tilde{\overline{w}}_r) \\ & + \sum_{r=1}^{N}\dot{f}_r\left(q\tilde{\underline{w}}_r + (1-q)\tilde{\overline{w}}_r\right)\end{aligned}$$

$$
\begin{aligned}
&= -\alpha \operatorname{sgn}(e) \\
&\quad + \sum_{r=1}^{N} \Bigg[\Big(\sum_{i=1}^{I} (\dot{a}_{ri} x_i + a_{ri} \dot{x}_i) + \dot{b}_r \Big) \\
&\quad \Big(q\underline{\tilde{w}}_r + (1-q)\tilde{\overline{w}}_r \Big) + f_r \Delta \underline{\tilde{w}}_r + f_r \Delta \tilde{\overline{w}}_r \Bigg]
\end{aligned} \tag{7.72}
$$

If (7.72) is inserted into the candidate Lyapunov function, (7.73) can be obtained:

$$
\begin{aligned}
\dot{V} &= \dot{e}e + \frac{1}{\gamma}\dot{\alpha}(\alpha - \alpha^*) = e(\dot{y}_N - \dot{y}) + \frac{1}{\gamma}\dot{\alpha}(\alpha - \alpha^*) \\
&= e\Bigg[-\alpha \operatorname{sgn}(e) + \sum_{r=1}^{N} \Bigg[\Big(\sum_{i=1}^{I} (\dot{a}_{ri} x_i + a_{ri} \dot{x}_i) + \dot{b}_r \Big) \times \\
&\quad \Big(q\underline{\tilde{w}}_r + (1-q)\tilde{\overline{w}}_r \Big) + f_r \Delta \underline{\tilde{w}}_r + f_r \Delta \tilde{\overline{w}}_r \Bigg] - \dot{y} \Bigg] + \frac{1}{\gamma}\dot{\alpha}(\alpha - \alpha^*)
\end{aligned} \tag{7.73}
$$

Following some steps similar to the previous section, we obtain the following equation:

$$
\begin{aligned}
\dot{V} &= -\mid e \mid 2\alpha + e\Bigg[\sum_{r=1}^{N} \Bigg[\sum_{i=1}^{I} \Bigg(-\alpha \operatorname{sgn}(e)\, x_i^2 \\
&\quad + a_{ri}\dot{x}_i \Big(q\underline{\tilde{w}}_r + (1-q)\tilde{\overline{w}}_r \Big) \Bigg) \Bigg] - \dot{y} + f_r \Delta \underline{\tilde{w}}_r + f_r \Delta \tilde{\overline{w}}_r \Bigg] + \frac{1}{\gamma}\dot{\alpha}(\alpha - \alpha^*)
\end{aligned}
$$

$$
\begin{aligned}
\dot{V} &< -\mid e \mid 2\alpha + \mid e \mid \Bigg[\sum_{r=1}^{N} \Bigg[\sum_{i=1}^{I} \Bigg(-\alpha B_{x^2} \\
&\quad + B_a B_{\dot{x}} \Big(q\underline{\tilde{w}}_r + (1-q)\tilde{\overline{w}}_r \Big) \Bigg) \Bigg] + B_{\dot{y}} B_f B_{\Delta_w} \Bigg] + \frac{1}{\gamma}\dot{\alpha}(\alpha - \alpha^*)
\end{aligned}
$$

Considering the upper bounds of x_i's, $\dot{x}_i$, and a_{ri}'s as in the assumptions of Theorem 7.2, we have:

$$
\begin{aligned}
\dot{V} &< -\mid e \mid 2\alpha + \mid e \mid \Bigg[\sum_{r=1}^{N} \Bigg[-I\alpha B_{x^2} + IB_a B_{\dot{x}} \Big(q\underline{\tilde{w}}_r \\
&\quad + (1-q)\tilde{\overline{w}}_r \Big) \Bigg] + B_{\dot{y}} + B_f B_{\Delta_w} \Bigg] + \frac{1}{\gamma}\dot{\alpha}(\alpha - \alpha^*)
\end{aligned}
$$

$$< -|e|2\alpha + |e|\left[-I\alpha B_{x^2} + IB_a B_{\dot{x}}\left(q\sum_{r=1}^{N}\underline{\tilde{w}}_r\right.\right.$$
$$\left.\left. + (1-q)\sum_{r=1}^{N}\tilde{\overline{w}}_r\right) + B_{\dot{y}}\right] + \frac{1}{\gamma}\dot{\alpha}(\alpha - \alpha^*)$$

So that:

$$\dot{V} < -|e|2\alpha + |e|\left[-I\alpha B_{x^2} + IB_a B_{\dot{x}} + B_{\dot{y}} + B_f B_{\Delta_w}\right]$$
$$+ \frac{1}{\gamma}\dot{\alpha}(\alpha - \alpha^*)$$

and further:

$$\dot{V} < -|e|\left(2\alpha + I\alpha B_{x^2}\right) + |e|\left[IB_a B_{\dot{x}} + B_{\dot{y}} + B_f B_{\Delta_w}\right]$$
$$+ \frac{1}{\gamma}\dot{\alpha}(\alpha - \alpha^*)$$

The parameter α^* is considered to satisfy the following inequality:

$$\alpha^* \geq \frac{2(IB_a B_{\dot{x}} + B_{\dot{y}} B_f B_{\Delta_w})}{2 + IB_{x^2}}$$

where this parameter is considered as determined during the adaptation of the learning rate.

$$\dot{V} \leq -|e|(2\alpha + I\alpha B_{x^2}) + |e|(IB_a B_{\dot{x}} + B_{\dot{y}} + B_f B_{\Delta_w})$$
$$+ (2 + IB_{x^2})\alpha^*|e| - (2 + IB_{x^2})\alpha^*|e| + \frac{1}{\gamma}(\alpha - \alpha^*)\dot{\alpha} \quad (7.74)$$

so that:

$$\dot{V} \leq |e|(IB_a B_{\dot{x}} + B_{\dot{y}} + B_f B_{\Delta_w}) - (2 + IB_{x^2})\alpha^*|e|$$
$$- (2 + IB_{x^2})(\alpha^* - \alpha)|e| + \frac{1}{\gamma}(\alpha - \alpha^*)\dot{\alpha} \quad (7.75)$$

and further:

$$\dot{V} \leq | e | (IB_a B_{\dot{x}} + B_{\dot{y}} + B_f B_{\Delta_w}) - (2 + IB_{x^2})\alpha^* | e | + (\alpha^* - \alpha)\left[(2 + IB_{x^2}) | e | - \frac{1}{\gamma}\dot{\alpha} \right] \tag{7.76}$$

Using the adaptation law for the adaptive learning rate (α) as:

$$\dot{\alpha} = (2 + IB_{x^2})\gamma | e | -\nu\gamma\alpha \tag{7.77}$$

in which ν has a small real value. Using this adaptation law, the time derivative of the Lyapunov function can be rewritten as:

$$\dot{V} \leq | e | (IB_a B_{\dot{x}} + B_{\dot{y}} + B_f B_{\Delta_w}) - (2 + IB_{x^2})\alpha^* | e | + (\alpha^* - \alpha)\nu\alpha \tag{7.78}$$

so that:

$$\dot{V} \leq | e | (IB_a B_{\dot{x}} + B_{\dot{y}} + B_f B_{\Delta_w}) - (2 + IB_{x^2})\alpha^* | e | -\nu(\alpha - \frac{\alpha^*}{2})^2 + \frac{\nu\alpha^{*2}}{4} \tag{7.79}$$

Considering the fact that $\alpha^* \geq \frac{2(IB_a B_{\dot{x}}+B_{\dot{y}}+B_f B_{\Delta_w})}{2+IB_{x^2}}$, we have $| e | (IB_a B_{\dot{x}} + B_{\dot{y}} + B_f B_{\Delta_w}) - \frac{\alpha^*}{2}(2 + IB_{x^2}) | e |\leq 0$ and consequently:

$$\dot{V} \leq -\frac{\alpha^*}{2}(2 + IB_{x^2}) | e | +\frac{\nu\alpha^{*2}}{4} \tag{7.80}$$

So that error converges to a very small region around zero in which $| e |\leq \frac{\alpha^*\nu}{2(2+IB_{x^2})}$.

7.3 CONTROLLER DESIGN

7.3.1 Control Scheme Incorporating a T2FNN Structure

In this section, the control scheme that benefits from a T2FNN whose parameters are adapted by a SMC theory-based approach is presented. Figure 7.1 shows the control scheme in which a PD controller works in parallel with a T2FNN. The PD controller is there to guarantee the global

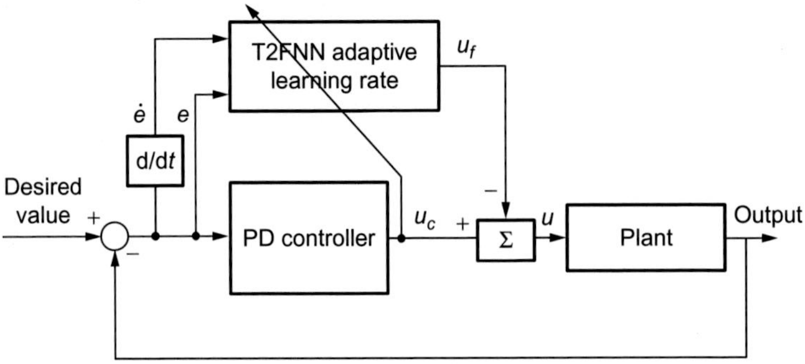

Figure 7.1 Structure of the proposed controller.

asymptotic stability in compact space and establishes a sliding manifold for the states of the system. The PD control signal is as follows:

$$u_c = k_P e + k_D \dot{e} \tag{7.81}$$

in which e is defined as $e = x_d - x$ and represents the error of the system, x_d is the reference input, and k_P, k_D are the gains of the PD controller.

7.3.1.1 T2FNN

A2-C0 fuzzy system is preferred as the structure of the fuzzy system. All the models have their closed forms to enable their use for further mathematical analysis. The ijth rule of a A2-C0 fuzzy system with two inputs can be written as:

$$\text{IF } x_1 \text{ is } \tilde{A}_{1i} \text{ and } x_2 \text{ is } \tilde{A}_{2j} \text{ THEN } u_f = f_{ij} \quad i = 1, .., I\, j = 1, .., J \tag{7.82}$$

where x_1 and x_2 are the input variables, I and J are the number of MFs for x_1 and x_2, respectively, u_f is the output variable, f_{ij}'s are the parameters in the consequent part of the fuzzy system, and $\tilde{A}_{1i}, \tilde{A}_{2j}$ are the type-2 fuzzy sets for the first and second input and their corresponding fuzzy MFs are $\tilde{\mu}_{1i}, \tilde{\mu}_{2j}$, respectively. The upper and lower MFs of $\tilde{\mu}_{1i}$ are considered to be $\overline{\mu}_{1i}$ and $\underline{\mu}_{1i}$ and the upper and lower MFs of $\tilde{\mu}_{2j}$ are considered to be $\overline{\mu}_{2j}$ and $\underline{\mu}_{2j}$. The final output of the system can be written as [8]:

$$y = q\frac{\sum_{j=1}^{J}\sum_{i=1}^{I}\underline{W}_{ij}f_{ij}}{\sum_{j=1}^{J}\sum_{i=1}^{I}\underline{W}_{ij}} + (1-q)\frac{\sum_{j=1}^{J}\sum_{i=1}^{I}\overline{W}_{ij}f_{ij}}{\sum_{j=1}^{J}\sum_{i=1}^{I}\overline{W}_{ij}} \tag{7.83}$$

where $\underline{W}_{ij} = \underline{\mu}_{1i}\underline{\mu}_{2j}$, $\overline{W}_{ij} = \overline{\mu}_{1i}\overline{\mu}_{2j}$, and the parameter q is the weighting parameter that reflects the sharing of the contribution of the upper and lower MFs.

Type-2 MFs, which are used for the control purposes, can be Gaussian type-2 MF with uncertain σ value or elliptic type-2 MF.

The A2-C0 fuzzy system that benefits from type-2 MFs in the premise part and crisp numbers for the consequent part is used for control purposes. Different from the identification part in which first order TSK fuzzy system is used, in the control part its zero order variant is used. The fuzzy *if-then* rule R_{ij} of a zero-order TSK model with two input variables where the consequent part has a constant value is defined as follows:

$$R_{ij} : \textit{ If } e \textit{ is } \tilde{A}_{1i,k} \text{ and } \dot{e} \textit{ is } \tilde{A}_{2j}, \text{ then } u_f = f_{ij}$$

where f_{ij} are constant values that are updated during the training. Using uncertain values for σ the type-2 MF has a footprint of uncertainty that is bounded with an upper and lower MF. The upper and lower MFs are denoted as $\overline{\mu}(x)$ and $\underline{\mu}(x)$, respectively. The firing strength of the rule R_{ij} is obtained as a T-norm of the MFs in the premise part (by using a multiplication operator):

$$\underline{W}_{ij} = \underline{\mu}_{1i}(e)\underline{\mu}_{2j}(\dot{e}) \tag{7.84}$$

$$\overline{W}_{ij} = \overline{\mu}_{1i}(e)\overline{\mu}_{2j}(\dot{e}) \tag{7.85}$$

where the real constants $\underline{\sigma}$, $\overline{\sigma} > 0$, and c are among the tunable parameters of the above T2FNN structure.

Hence, (7.84) and (7.85) can be rewritten as follows:

$$\underline{W}_{ij} = \exp\left[-(\frac{e - c_{1i}}{\underline{\sigma}_{1i}})^2 - (\frac{\dot{e} - c_{2j}}{\underline{\sigma}_{2j}})^2\right] \tag{7.86}$$

$$\overline{W}_{ij} = \exp\left[-(\frac{e - c_{1i}}{\overline{\sigma}_{1i}})^2 - (\frac{\dot{e} - c_{2j}}{\overline{\sigma}_{2j}})^2\right] \tag{7.87}$$

The computational output of A2-C0 structure is derived as:

$$u_f = \frac{q\sum_{i=1}^{I}\sum_{j=1}^{J} f_{ij}\underline{W}_{ij}}{\sum_{i=1}^{I}\sum_{j=1}^{J}\underline{W}_{ij}} + \frac{(1-q)\sum_{i=1}^{I}\sum_{j=1}^{J} f_{ij}\overline{W}_{ij}}{\sum_{i=1}^{I}\sum_{j=1}^{J}\overline{W}_{ij}} \tag{7.88}$$

After the normalization of (7.88), the output signal of the T2FNN will acquire the following form:

$$u_f = q\sum_{i=1}^{I}\sum_{j=1}^{J} f_{ij}\underline{\widetilde{W}}_{ij} + (1-q)\sum_{i=1}^{I}\sum_{j=1}^{J} f_{ij}\widetilde{\overline{W}}_{ij} \tag{7.89}$$

where $\underline{\widetilde{W}}_{ij}$ and $\widetilde{\overline{W}}_{ij}$ are the normalized values of the lower and upper outputs corresponding to the ijth rule of the fuzzy system:

$$\underline{\widetilde{W}}_{ij} = \frac{\underline{W}_{ij}}{\sum_{i=1}^{I}\sum_{j=1}^{J}\underline{W}_{ij}} \tag{7.90}$$

$$\widetilde{\overline{W}}_{ij} = \frac{\overline{W}_{ij}}{\sum_{i=1}^{I}\sum_{j=1}^{J}\overline{W}_{ij}} \tag{7.91}$$

$\underline{\widetilde{W}}(t) = \left[\underline{\widetilde{W}}_{11}(t)\ \underline{\widetilde{W}}_{12}(t)\ldots\ \underline{\widetilde{W}}_{21}(t)\ \ldots\ \underline{\widetilde{W}}_{ij}(t)\ \ldots\ \underline{\widetilde{W}}_{IJ}(t)\right]^{\mathrm{T}}$ are the vector of the normalized lower output signals of the neurons from the second hidden layer of the first and second T2FNNs, respectively; $\widetilde{\overline{W}}(t) = \left[\widetilde{\overline{W}}_{11}(t)\ \widetilde{\overline{W}}_{12}\ (t)\ldots\ \widetilde{\overline{W}}_{21}(t)\ \ldots\ \widetilde{\overline{W}}_{ij}(t)\ \ldots\ \widetilde{\overline{W}}_{IJ}(t)\right]^{\mathrm{T}}$ is the vector of the normalized upper output signals of the neurons from the second hidden layer of the IT2FNNs; and $\underline{\sigma}_1 = \left[\underline{\sigma}_{11}\ \ldots\ \underline{\sigma}_{1i}\ \ldots\ \underline{\sigma}_{1I}\right]^{\mathrm{T}}$, $\underline{\sigma}_2 = \left[\underline{\sigma}_{21}\ \ldots\ \underline{\sigma}_{2j}\ \ldots\ \underline{\sigma}_{2J}\right]^{\mathrm{T}}$ are vectors of the tuning parameters $\underline{\sigma}$ of the lower bounds of type-2 Gaussian MFs with uncertain variance. $\overline{\sigma}_1 = [\overline{\sigma}_{11}\ \ldots\ \overline{\sigma}_{1i}\ \ldots\ \overline{\sigma}_{1I}]^{\mathrm{T}}$, $\overline{\sigma}_2 = \left[\overline{\sigma}_{21}\ \ldots\ \overline{\sigma}_{2j}\ \ldots\ \overline{\sigma}_{2J}\right]^{\mathrm{T}}$ are vectors of the tuning parameters $\overline{\sigma}$ of the upper bounds of type-2 Gaussian MFs with uncertain variance. $c_1 = [c_{11}\ \ldots\ c_{1i}\ \ldots\ c_{1I}]^{\mathrm{T}}$ and $c_2 = \left[c_{21}\ \ldots\ c_{2j}\ \ldots\ c_{2J}\right]^{\mathrm{T}}$ are vectors of the tuning parameters c of the lower and upper bounds of type-2 Gaussian MFs with uncertain variance.

It is assumed that due to the control scheme used in Fig. 7.1, where the conventional controller serves to guarantee global asymptotic stability in compact space, the input signals, $e(t)$ and $\dot{e}(t)$, and their time derivatives can be considered bounded:

$$|e(t)| \le B_e,\ \ |\dot{e}(t)| \le B_{\dot{e}}\ |\ddot{e}(t)| \le B_{\ddot{e}} \quad \forall t \tag{7.92}$$

where B_e and $B_{\dot{e}}$ are assumed to be some unknown positive constants. The adaptation law for the tunable parameters is considered such that $\underline{\sigma}$, $\overline{\sigma}$, and c of the Gaussian MFs are bounded as follows:

$$\underline{B}_\sigma \leq \left\|\underline{\sigma}_1\right\| \leq \overline{B}_\sigma,\ \underline{B}_\sigma \leq \left\|\underline{\sigma}_2\right\| \leq \overline{B}_\sigma, \tag{7.93}$$
$$\|c_1\| \leq B_c,\quad \|c_2\| \leq B_c$$
$$\underline{B}_\sigma \leq \|\overline{\sigma}_1\| \leq \overline{B}_\sigma,\ \underline{B}_\sigma \leq \|\overline{\sigma}_2\| \leq \overline{B}_\sigma, \tag{7.94}$$
$$\|\overline{c}_1\| \leq B_c,\quad \|\overline{c}_2\| \leq B_c \left\|\overline{f}_{ij}\right\| \leq B_f$$

where $\underline{B}_\sigma$, $\overline{B}_\sigma$, B_c, and B_f are some unknown positive constants.

Similar to the previous case, it follows that $0 < \underline{\widetilde{W}}_{ij} < 1$ and $0 < \widetilde{\widetilde{W}}_{ij} < 1$. In addition, by definition, $\sum_{i=1}^{I}\sum_{j=1}^{J}\underline{\widetilde{W}}_{ij} = 1$ and $\sum_{i=1}^{I}\sum_{j=1}^{J}\widetilde{\widetilde{W}}_{ij} = 1$. It is also considered that u and $\dot{u}$ are also bounded signals, i.e.,

$$|u(t)| \leq B_u,\quad |\dot{u}(t)| \leq B_{\dot{u}}\quad \forall t \tag{7.95}$$

where B_u and $B_{\dot{u}}$ are two unknown positive constants.

7.3.2 SMC Theory-Based Learning Algorithm for T2FNN with Gaussian MFs with Uncertain σ Values

The zero value of the learning error coordinate $u_c(t)$ can be defined [9] as time-varying sliding surface, i.e.,

$$S_c\left(u_f, u\right) = u_c(t) = u_f(t) + u(t) \tag{7.96}$$

The sliding manifold works as a guideline to train the parameters of T2FNN. This sliding surface acts as the condition that the T2FNN is trained to become a nonlinear regulator to obtain the desired response during the tracking-error convergence by compensating the nonlinearity of the controlled plant. The sliding surface for the nonlinear system under control $S_p(e, \dot{e})$ is defined as:

$$S_p(e, \dot{e}) = \dot{e} + \chi e \tag{7.97}$$

with χ being a constant determining the slope of the sliding surface.

Definition: An sliding motion will appear on the sliding line $S_c\left(u_f, u\right) = u_c(t) = 0$ after a time t_h, if the condition $S_c(t)\dot{S}_c(t) = u_c(t)\,\dot{u}_c(t) < 0$ is satisfied for all t in some nontrivial semi-open subinterval of time of the form $[t, t_h) \subset (-\infty, t_h)$.

It is best to design an online learning algorithm for the parameters of T2FNN such that the sliding mode condition of the above definition is enforced.

7.3.2.1 Parameter Update Rules For T2FNN

Theorem 7.3. *If the adaptation laws for T2FNN parameters are chosen as:*

$$\dot{c}_{1i} = -\beta_1 \frac{\overline{\sigma}_{1i}^2}{e - c_{1i}} \operatorname{sgn}(u_c) \tag{7.98}$$

$$\dot{c}_{2j} = -\beta_1 \frac{\overline{\sigma}_{2j}^2}{\dot{e} - c_{2j}} \operatorname{sgn}(u_c) \tag{7.99}$$

$$\dot{\underline{\sigma}}_{1i} = -\beta_1 \frac{\underline{\sigma}_{1i}^3}{(e - c_{1i})^2} \operatorname{sgn}(u_c) \tag{7.100}$$

$$\dot{\underline{\sigma}}_{2j} = -\beta_1 \frac{\underline{\sigma}_{2j}^3}{(\dot{e} - c_{2j})^2} \operatorname{sgn}(u_c) \tag{7.101}$$

$$\dot{\overline{\sigma}}_{1i} = -\beta_1 \frac{\overline{\sigma}_{1i}^3}{(e - c_{1i})^2} \operatorname{sgn}(u_c) \tag{7.102}$$

$$\dot{\overline{\sigma}}_{2j} = -\beta_1 \frac{(\overline{\sigma}_{2j})^3}{(\dot{e} - c_{2j})^2} \operatorname{sgn}(u_c) \tag{7.103}$$

$$\dot{f}_{ij} = -\alpha \frac{q\underline{\widetilde{W}}_{ij} + (1-q)\overline{\widetilde{W}}_{ij}}{\left(q\underline{\widetilde{W}} + (1-q)\overline{\widetilde{W}}\right)^T \left(q\underline{\widetilde{W}} + (1-q)\overline{\widetilde{W}}\right)} sgn(u_c) \tag{7.104}$$

$$\dot{\alpha} = \gamma_1 |u_c| - \nu\gamma_1\alpha \tag{7.105}$$

then, given an arbitrary initial condition $u_c(0)$, the learning error $u_c(t)$ will converge firmly to zero during a finite time t_h.

There is a relation between the sliding line S_p and $u_c(t)$ (the classical controller signal). If χ is taken as $\chi = \frac{k_P}{k_D}$, the following equation is obtained:

$$S_c = u_c = k_D\dot{e} + k_P e = k_D\left(\dot{e} + \frac{k_P}{k_D}e\right) = k_D S_p \tag{7.106}$$

Remark. Equation (7.106) indicates that the convergence of S_c toward zero guarantees the convergence of S_p toward zero and there exists a sliding motion in the states of the system.

7.3.2.2 Proof of Theorem 7.3

The stability analysis of the learning algorithm is considered in this section. The following variables are defined:

$$A_{1i} = \left[-(\frac{e - c_{1i}}{\underline{\sigma}_{1i}})^2 \right] \tag{7.107}$$

$$U_{1i} = \left[-(\frac{e - c_{1i}}{\overline{\sigma}_{1i}})^2 \right] \tag{7.108}$$

$$A_{2j} = \left[-(\frac{\dot{e} - c_{2j}}{\underline{\sigma}_{2j}})^2 \right] \tag{7.109}$$

$$U_{2j} = \left[-(\frac{\dot{e} - c_{2j}}{\overline{\sigma}_{2j}})^2 \right] \tag{7.110}$$

Considering (7.107)-(7.110) we have:

$$\dot{\underline{\mu}}_{1i}(e) = -2A_{1i}\dot{A}_{1i}\underline{\mu}_{1i}(e) \tag{7.111}$$

$$\dot{\overline{\mu}}_{1i}(e) = -2U_{1i}\dot{U}_{1i}\overline{\mu}_{1i}(e) \tag{7.112}$$

$$\dot{\underline{\mu}}_{2j}(\dot{e}) = -2A_{2j}(\dot{A}_{2j})\underline{\mu}_{2j}(\dot{e}) \tag{7.113}$$

$$\dot{\overline{\mu}}_{2j}(\dot{e}) = -2U_{2j}(\dot{U}_{2j})\overline{\mu}_{2j}(\dot{e}) \tag{7.114}$$

In addition, considering (7.90) and (7.91), the time derivatives of $\underline{\widetilde{W}}_{ij}$ and $\overline{\widetilde{W}}_{ij}$ are obtained as:

$$\underline{\widetilde{W}}_{ij} = \frac{\underline{W}_{ij}}{\sum_{i=1}^{I}\sum_{j=1}^{J}\underline{W}_{ij}} \Rightarrow \dot{\underline{\widetilde{W}}}_{ij} = \frac{\left(\underline{\mu}_{1i}(e)\underline{\mu}_{2j}(\dot{e})\right)'\left(\sum_{i=1}^{I}\sum_{j=1}^{J}\underline{W}_{ij}\right)}{\left(\sum_{i=1}^{I}\sum_{j=1}^{J}\underline{W}_{ij}\right)^2} - \frac{\underline{W}_{ij}\left(\sum_{i=1}^{I}\sum_{j=1}^{J}\underline{\mu}_{1i}(e)\underline{\mu}_{2j}(\dot{e})\right)'}{\left(\sum_{i=1}^{I}\sum_{j=1}^{J}\underline{W}_{ij}\right)^2} \tag{7.115}$$

$$\widetilde{\overline{W}}_{ij} = \frac{\overline{W}_{ij}}{\sum_{i=1}^{I}\sum_{j=1}^{J}\overline{W}_{ij}} \Rightarrow \dot{\widetilde{\overline{W}}}_{ij} = \frac{\left(\overline{\mu}_{1i}(e)\overline{\mu}_{2j}(\dot{e})\right)'\left(\sum_{i=1}^{I}\sum_{j=1}^{J}\overline{W}_{ij}\right)}{\left(\sum_{i=1}^{I}\sum_{j=1}^{J}\overline{W}_{ij}\right)^2} - \frac{\overline{W}_{ij}\left(\sum_{i=1}^{I}\sum_{j=1}^{J}\overline{\mu}_{1i}(e)\overline{\mu}_{2j}(\dot{e})\right)'}{\left(\sum_{i=1}^{I}\sum_{j=1}^{J}\overline{W}_{ij}\right)^2} \tag{7.116}$$

Since $\underline{\tilde{W}}_{ij} = (\underline{W}_{ij})/(\sum_{i=1}^{I}\sum_{j=1}^{J}\underline{W}_{ij})$ and $\tilde{\overline{W}}_{ij} = (\overline{W}_{ij})/(\sum_{i=1}^{I}\sum_{j=1}^{J}\overline{W}_{ij})$, we have:

$$\begin{aligned}
\dot{\underline{\widetilde{W}}}_{ij} &= \frac{\dot{\underline{\mu}}_{1i}(e)\underline{\mu}_{2j}(\dot{e}) + \underline{\mu}_{1i}(e)\dot{\underline{\mu}}_{2j}(\dot{e})}{\sum_{i=1}^{I}\sum_{j=1}^{J}\underline{W}_{ij}} - \frac{\underline{\widetilde{W}}_{ij}\sum_{i=1}^{I}\sum_{j=1}^{J}\left(\dot{\underline{\mu}}_{1i}(e)\underline{\mu}_{2j}(\dot{e}) + \underline{\mu}_{1i}(e)\dot{\underline{\mu}}_{2j}(\dot{e})\right)}{\sum_{i=1}^{I}\sum_{j=1}^{J}\underline{W}_{ij}} \\
&= \frac{-2A_{1i}\dot{A}_{1i}\underline{\mu}_{1i}(e)\underline{\mu}_{2j}(\dot{e}) - 2A_{2j}\dot{A}_{2j}\underline{\mu}_{1i}(e)\underline{\mu}_{2j}(\dot{e})}{\sum_{i=1}^{I}\sum_{j=1}^{J}\underline{W}_{ij}} - \frac{\underline{\widetilde{W}}_{ij}\sum_{i=1}^{I}\sum_{j=1}^{J}\left(-2A_{2j}\dot{A}_{2j}\underline{\mu}_{1i}(e)\underline{\mu}_{2j}(\dot{e})\right)}{\sum_{i=1}^{I}\sum_{j=1}^{J}\underline{W}_{ij}} \\
&\quad - \frac{\underline{\widetilde{W}}_{ij}\sum_{i=1}^{I}\sum_{j=1}^{J}\left(-2A_{1i}\dot{A}_{1i}\underline{\mu}_{1i}(e)\underline{\mu}_{2j}(\dot{e})\right)}{\sum_{i=1}^{I}\sum_{j=1}^{J}\underline{W}_{ij}}
\end{aligned} \tag{7.117}$$

and further:

$$\dot{\underline{\widetilde{W}}}_{ij} = -\underline{\widetilde{W}}_{ij}\dot{\underline{N}}_{ij} + \underline{\widetilde{W}}_{ij}\sum_{i=1}^{I}\sum_{i=1}^{J}\underline{\widetilde{W}}_{ij}\dot{\underline{N}}_{ij} \tag{7.118}$$

$$\dot{\widetilde{\overline{W}}}_{ij} = -\widetilde{\overline{W}}_{ij}\dot{\overline{N}}_{ij} + \widetilde{\overline{W}}_{ij}\sum_{i=1}^{I}\sum_{i=1}^{J}\widetilde{\overline{W}}_{ij}\dot{\overline{N}}_{ij} \tag{7.119}$$

in which:

$$\dot{\underline{N}}_{ij} = 2A_{1i}\dot{A}_{1i} + 2A_{2j}\dot{A}_{2j},\ \dot{\overline{N}}_{ij} = 2U_{1i}\dot{U}_{1i} + 2U_{2j}\dot{U}_{2j} \tag{7.120}$$

$$\dot{A}_{1i} = \frac{(\dot{e} - \dot{c}_{1i})\underline{\sigma}_{1i} - (e - \underline{c}_{1i})\dot{\underline{\sigma}}_{1i}}{\underline{\sigma}_{1i}^2}, \dot{A}_{2j} = \frac{(\ddot{e} - \dot{\underline{c}}_{2j})\underline{\sigma}_{2j} - (\dot{e}_k - \underline{c}_{2j})\dot{\underline{\sigma}}_{2j}}{\underline{\sigma}_{2j}^2}$$

$$\dot{U}_{1i} = \frac{(\dot{e} - \dot{\overline{c}}_{1i})\overline{\sigma}_{1i} - (e - \overline{c}_{1i})\dot{\overline{\sigma}}_{1i}}{\overline{\sigma}_{1i}^2}, \dot{U}_{2j} = \frac{(\ddot{e} - \dot{\overline{c}}_{2j})\overline{\sigma}_{2j} - (\dot{e} - \overline{c}_{2j})\dot{\overline{\sigma}}_{2j}}{\overline{\sigma}_{2j}^2}$$

It is possible to use Maclaurin series expansion to obtain following equations:

$$\begin{aligned} \frac{\dot{x}_1 - \dot{c}_{1i}}{\underline{\sigma}_{1i}} &= \frac{\dot{e} - \dot{c}_{1i}}{\underline{\sigma}_{1i} - \overline{\sigma}_{1i} + \overline{\sigma}_{1i}} \\ &= \frac{\dot{e} - \dot{c}_{1i}}{\overline{\sigma}_{1i}} \left(1 - \underbrace{\frac{\underline{\sigma}_{1i} - \overline{\sigma}_{1i}}{\overline{\sigma}_{1i}} + \frac{(\overline{\sigma}_{1i} - \underline{\sigma}_{1i})^2}{\underline{\sigma}_{1i}^2} + H.O.T}_{D_{1i}} \right) \end{aligned} \tag{7.121}$$

and:

$$\begin{aligned} \frac{e - c_{1i}}{\underline{\sigma}_{1i}} &= \frac{e - c_{1i}}{\underline{\sigma}_{1i} - \overline{\sigma}_{1i} + \overline{\sigma}_{1i}} \\ &= \frac{e - c_{1i}}{\overline{\sigma}_{1i}} \left(1 - \frac{\underline{\sigma}_{1i} - \overline{\sigma}_{1i}}{\overline{\sigma}_{1i}} + \frac{(\overline{\sigma}_{1i} - \underline{\sigma}_{1i})^2}{\underline{\sigma}_{1i}} + H.O.T \right) \end{aligned} \tag{7.122}$$

Using (7.121), we have:

$$A_{1i}\dot{A}_{1i} = U_{1i}\dot{U}_{1i} + \frac{(\dot{e} - \dot{c}_{1i})}{\overline{\sigma}_{1i}} \frac{(e - c_{1i})}{\overline{\sigma}_{1i}} \left(2D_{1i} + D_{1i}^2 \right) \tag{7.123}$$

and similarly using (7.122), we find that:

$$A_{2j}\dot{A}_{2j} = U_{2j}\dot{U}_{2j} + \frac{(\ddot{e} - \dot{c}_{2j})}{\overline{\sigma}_{2j}} \frac{(\dot{e} - c_{2j})}{\overline{\sigma}_{2j}} \left(2E_{2j} + E_{2j}E_{2j}^2 \right) \tag{7.124}$$

in which:

$$E_{2j} = -\frac{\underline{\sigma}_{2j} - \overline{\sigma}_{2j}}{\overline{\sigma}_{2j}} + \frac{(\underline{\sigma}_{2j} - \overline{\sigma}_{2j})^2}{\overline{\sigma}_{2j}^2} + H.O.T \tag{7.125}$$

It can be proved that $|D_{1i}|$ and $|E_{2j}|$ are bounded as $|D_{1i}| < B_D$ and $|E_{2j}| < B_E$. In order to analyze the stability of the controller with an adaptive learning rate, the following Lyapunov function is proposed:

$$V_c = \frac{1}{2}u_c^2(t) + \frac{1}{2\gamma_1}(\alpha - \alpha^*)^2 \tag{7.126}$$

The time derivative of the Lyapunov function (7.126) is derived as:

$$\dot{V}_c = u_c \dot{u}_c = u_c(\dot{u}_f + \dot{u}) + \frac{\dot{\alpha}}{\gamma_1}(\alpha - \alpha^*) \tag{7.127}$$

Since:

$$\begin{aligned} u_f &= \frac{q\sum_{i=1}^{I}\sum_{j=1}^{J} f_{ij}\underline{W}_{ij}}{\sum_{i=1}^{I}\sum_{j=1}^{J}\underline{W}_{ij}} + \frac{(1-q)\sum_{i=1}^{I}\sum_{j=1}^{J} f_{ij}\overline{W}_{ij}}{\sum_{i=1}^{I}\sum_{j=1}^{J}\overline{W}_{ij}} \\ &= q\sum_{i=1}^{I}\sum_{j=1}^{J} f_{ij}\underline{\widetilde{W}}_{ij} + (1-q)\sum_{i=1}^{I}\sum_{j=1}^{J} f_{ij}\overline{\widetilde{W}}_{ij} \end{aligned} \tag{7.128}$$

and:

$$\dot{u}_f = q\sum_{i=1}^{I}\sum_{j=1}^{J}(\dot{f}_{ij}\underline{\tilde{W}}_{ij} + f_{ij}\dot{\underline{\tilde{W}}}_{ij}) + (1-q)\sum_{i=1}^{I}\sum_{j=1}^{J}(\dot{f}_{ij}\overline{\widetilde{W}}_{ij} + f_{ij}\dot{\overline{\widetilde{W}}}_{ij}) \tag{7.129}$$

we obtain:

$$\begin{aligned} \dot{u}_f = q\sum_{i=1}^{I}\sum_{j=1}^{J}\left(\left(-\underline{\tilde{W}}_{ij}\dot{\underline{K}}_{ij} + \underline{\tilde{W}}_{ij}\sum_{i=1}^{I}\sum_{j=1}^{J}\underline{\tilde{W}}_{ij}\dot{\underline{K}}_{ij}\right)f_{ij} + \underline{\tilde{W}}_{ij}\dot{f}_{ij}\right) \\ + (1-q)\sum_{i=1}^{I}\sum_{j=1}^{J}\left(\left(-\overline{\widetilde{W}}_{ij}\dot{\overline{K}}_{ij} + \overline{\widetilde{W}}_{ij}\sum_{i=1}^{I}\sum_{j=1}^{J}\overline{\widetilde{W}}_{ij}\dot{\overline{K}}_{ij}\right)f_{ij} + \overline{\widetilde{W}}_{ij}\dot{f}_{ij}\right) \end{aligned} \tag{7.130}$$

Considering adaptation for sigma and center we have:

$$
\begin{aligned}
\dot{V}_c = u_c \Bigg(& q \sum_{i=1}^{I} \sum_{j=1}^{J} \Bigg(2 \Bigg(-\underline{\widetilde{W}}_{ij} \Big(\frac{\dot{e}}{\underline{\sigma}_{1i}} A_{1i} + \frac{\ddot{e}}{\underline{\sigma}_{2j}} A_{2j} \Big) \\
& + \underline{\widetilde{W}}_{ij} \sum_{i=1}^{I} \sum_{j=1}^{J} \underline{\widetilde{W}}_{ij} \Big(\frac{\dot{e}}{\underline{\sigma}_{1i}} A_{1i} + \frac{\ddot{e}}{\underline{\sigma}_{2j}} A_{2j} \Big) \Bigg) f_{ij} + \underline{\widetilde{W}}_{ij} \dot{f}_{ij} \Bigg) \\
& + q \sum_{i=1}^{I} \sum_{j=1}^{J} \Bigg(2 \Big(-\underline{\widetilde{W}}_{ij} \big(2D_{1i} + D_{1i}^2 + 2D_{2j} + D_{2j}^2 \big) \\
& + \underline{\widetilde{W}}_{ij} \sum_{i=1}^{I} \sum_{j=1}^{J} \underline{\widetilde{W}}_{ij} \big(2D_{1i} + D_{1i}^2 + 2D_{2j} + D_{2j}^2 \big) \Big) f_{ij} \Bigg) \\
& + (1-q) \sum_{i=1}^{I} \sum_{j=1}^{J} \Bigg(2 \Big(-\overline{\widetilde{W}}_{ij} \Big(\frac{\dot{e}}{\overline{\sigma}_{1i}} A_{1i} + \frac{\ddot{e}}{\overline{\sigma}_{2j}} A_{2j} \Big) \\
& + \overline{\widetilde{W}}_{ij} \sum_{i=1}^{I} \sum_{j=1}^{J} \overline{\widetilde{W}}_{ij} \Big(\frac{\dot{e}}{\overline{\sigma}_{1i}} A_{1i} + \frac{\ddot{e}}{\overline{\sigma}_{2j}} A_{2j} \Big) \Big) f_{ij} + \overline{\widetilde{W}}_{ij} \dot{f}_{ij} \Bigg) + \dot{u} \Bigg) \\
& + \frac{\dot{\alpha}}{\gamma_1} (\alpha - \alpha^*)
\end{aligned}
\tag{7.131}
$$

Considering assumptions (7.93) and (7.94), (7.131) can be rewritten as:

$$
\begin{aligned}
\dot{V}_c &\leq 4B_r |u_c| + u_c \left(\sum_{i=1}^{I} \sum_{j=1}^{J} \dot{f}_{ij} \big(q \underline{\widetilde{W}}_{ij} + (1-q) \overline{\widetilde{W}}_{ij} \big) + \dot{u} \right) + \frac{\dot{\alpha}}{\gamma_1} (\alpha - \alpha^*) \\
&\leq 4B_r |u_c| - \alpha^* |u_c| + (\alpha^* - \alpha) |u_c| + |u_c| B_{\dot{u}} + \frac{\dot{\alpha}}{\gamma_1} (\alpha - \alpha^*) < 0
\end{aligned}
\tag{7.132}
$$

in which:

$$
B_r = B_f \left(\frac{3B_D + 3B_E + B_e^2 + B_{\dot{e}}^2 + B_c B_e + B_c B_{\dot{e}}}{\underline{B}_{\sigma}^2} \right) \tag{7.133}
$$

using the adaptation law for α as:

$$\dot{\alpha} = \gamma_1 |u_c| - \nu\gamma_1\alpha$$

and taking α^* as:

$$B_{\ddot{u}} + 4B_r < \frac{1}{2}\alpha^*$$

we have:

$$\begin{aligned}\dot{V}_c &\leq -\frac{1}{2}\alpha^* |u_c| + \nu\alpha(\alpha - \alpha^*) \\ &= -\frac{1}{2}\alpha^* |u_c| + \nu(\alpha - \alpha^*)^2 - \frac{\nu\alpha^{*2}}{4}\end{aligned} \tag{7.134}$$

Furthermore:

$$\dot{V}_c \leq -\frac{1}{2}\alpha^* |u_c| + \nu(\alpha - \alpha^*)^2 \tag{7.135}$$

Therefore, the Lyapunov function u_c converges exponentially until $|u_c| \leq 2\frac{\nu}{\alpha^*}(\alpha - \alpha^*)^2$ and the parameters of the controller are bounded. Consequently, the system states converge to a compact set R in which:

$$R = \left\{u| \, |u| \leq 2\frac{\nu}{\alpha^*}(\alpha - \alpha^*)^2\right\} \tag{7.136}$$

It should be noted that this region can be chosen to be as small as desired by choosing an appropriate value for ν. Consequently, u can be made as small as desired.

7.3.3 SMC Theory-Based Learning Algorithm for T2FNN with Elliptic MFs

In this section, the SMC theory-based parameter update rules for the T2FNN with elliptic MFs are discussed. As will be seen later from the parameter update rules, the introduced novel training algorithm is simple and computationally less expensive than the gradient-based methods and the adaptation laws have closed forms. Although the gradient-based methods may result in instability, the stability of the proposed method is proved using the Lyapunov approach.

The sliding surface for the nonlinear system is defined as:

$$S_p(e, \dot{e}) = \dot{e} + \chi e \tag{7.137}$$

where χ is a positive variable that determines the desired trajectory of the error signal.

Definition: A sliding motion will appear on the sliding manifold $S_p(e, \dot{e}) = 0$ after a time t_h if the condition $S_p(t)\dot{S}_p(t) < 0$ is satisfied for all t's in some nontrivial semi-open subinterval of time of the form $[t, t_h) \subset (0, t_h)$.

Theorem 7.4. *Consider the control structure that consists of a PD controller that works in parallel with a T2FNN. The PD controller is responsible for guaranteeing the stability of the system to be controlled in the compact set and it further guarantees that:*

$$\begin{aligned} |f_{ij}| < B_f,\ |x_1| < B_x,\ |x_2| < B_x, \\ |\dot{x}_1| < B_{\dot{x}},\ |\dot{x}_2| < B_{\dot{x}} \\ B_{d,min} < |d_{1i}|,\ B_{d,min} < |d_{2j}|,\ |c_{1i}| < B_c, \\ |c_{2j}| < B_c,\ |u| < B_u \end{aligned} \tag{7.138}$$

Then, if the parameter update rules for the T2FNN are chosen as (7.139)-(7.149) *with the selection of the adaptive learning rate (α) as in* (7.149), *for given arbitrary initial conditions $u_c(0)$, $u_c(t)$ converges asymptotically to zero in finite time and sliding motion will be achieved.*

$$\dot{a}_{2,1i} = \gamma \left\{ \frac{ln\left(1 - \left|\frac{x_1 - c_{1i}}{d_{1i}}\right|^{a_{2,1i}}\right)}{a_{2,1i}^2} + \frac{\left|\frac{x_1 - c_{1i}}{d_{1i}}\right|^{a_{2,1i}} ln\left|\frac{x_1 - c_{1i}}{d_{1i}}\right|}{a_{2,1i}(1 - T_{2,1i})} \right\}^{-1} H(x_1, c_{1i}, d_{1i}) \tag{7.139}$$

$$\dot{a}_{1,1i} = \gamma \left\{ \frac{ln\left(1 - \left|\frac{x_1 - c_{1i}}{d_{1i}}\right|^{a_{1,1i}}\right)}{a_{1,1i}^2} + \frac{\left|\frac{x_1 - c_{1i}}{d_{1i}}\right|^{a_{1,1i}} ln\left|\frac{x_1 - c_{1i}}{d_{1i}}\right|}{a_{1,1i}(1 - T_{1,1i})} \right\}^{-1} H(x_1, c_{1i}, d_{1i}) \tag{7.140}$$

$$\dot{a}_{2,2j} = \gamma \left\{ \frac{ln\left(1 - \left|\frac{x_2 - c_{2j}}{d_{2j}}\right|^{a_{2,2j}}\right)}{a_{2,2j}^2} + \frac{\left|\frac{x_2 - c_{2j}}{d_{2j}}\right|^{a_{2,2j}} ln\left|\frac{x_2 - c_{2j}}{d_{2j}}\right|}{a_{2,2j}(1 - T_{2,2j})} \right\}^{-1} H(x_2, c_{2j}, d_{2j}) \tag{7.141}$$

$$\dot{a}_{1,2j} = \gamma \left\{ \frac{ln\left(1 - \left|\frac{x_2 - c_{2j}}{d_{2j}}\right|^{a_{1,2j}}\right)}{a_{1,2j}^2} + \frac{\left|\frac{x_2 - c_{2j}}{d_{2j}}\right|^{a_{1,2j}} ln\left|\frac{x_2 - c_{2j}}{d_{2j}}\right|}{a_{1,2j}(1 - T_{1,2j})} \right\}^{-1} H(x_2, c_{2j}, d_{2j}) \tag{7.142}$$

$$\dot{c}_{1i} = -\gamma \frac{|d_{1i}|^{a_{2,1i}}(1 - T_{2,1i})\mathrm{sgn}(x_1 - c_{1i})}{|x_1 - c_{1i}|^{a_{2,1i}-1}} \tag{7.143}$$

$$\dot{d}_{1i} = \gamma \frac{(1 - T_{2,1i})|d_{1i}|^{a_{2,1i}+1}}{|x_1 - c_{1i}|^{a_{2,1i}}} \mathrm{sgn}(d_{1i}) \tag{7.144}$$

$$\dot{c}_{2j} = -\gamma \frac{|d_{2j}|^{a_{2,2j}}(1 - T_{2,2j})\mathrm{sgn}(x_2 - c_{2j})}{|x_2 - c_{2j}|^{a_{2,2j}-1}} \tag{7.145}$$

$$\dot{d}_{2j} = \gamma \frac{(1 - T_{2,2j})|d_{1i}|^{a_{2,2j}+1}}{|x_2 - c_{2j}|^{a_{2,2j}}} \mathrm{sgn}(d_{2j}) \tag{7.146}$$

$$\dot{f}_{ij} = -\frac{q\underline{\widetilde{W}}_{ij} + (1 - q)\overline{\widetilde{W}}_{ij}}{\Pi_{ij}^{\mathrm{T}} \Pi_{ij}} \alpha \mathrm{sign}(u_c) \tag{7.147}$$

in which:

$$\Pi_{ij} = \left(\sum_{i=1}^{I} \sum_{j=1}^{J} (q\underline{\widetilde{W}}_{ij} + (1 - q)\overline{\widetilde{W}}_{ij}) \right) \tag{7.148}$$

$$\dot{\alpha} = 2\gamma_1 |u_c| \tag{7.149}$$

In the above, γ_1 and γ are the learning rates considered for the adaptation of the learning rate and they need to be selected as positive.

Remark 7.5. As can be seen from (7.149), an adaptive learning rate is considered for updating the parameters of the consequent parts. Thus, a priori knowledge about the upper bound of the states of the system is not needed.

If the system under control is a second-order dynamic system in the form of:

$$\begin{aligned} \dot{x}_1 &= x_2 \\ \dot{x}_2 &= f(x_1, x_2) + u \end{aligned} \tag{7.150}$$

then it is possible to extend the results of Theorem 7.1 and prove the stability of the system without any assumptions on any bounds on the states of the system. The following theorem summarizes these results.

Theorem 7.5. *Consider the control structure as shown in Fig. 7.1, which consists of a PD controller working in parallel with a T2FNN and a robustness term (i.e.,* $u = K_p e + K_d \dot{e} + u_r - u_f$*) and* $u_r = K_r sign(\dot{e}) + \ddot{x}_d$ *in which* $\dot{K}_r = \gamma |\dot{e}|,\ 0 < \gamma$*. In addition, the plant under control is a second-order dynamic system in the form of* (7.150)*. It is assumed that the function* $f(x_1, x_2)$ *is bounded in a compact set (i.e.,* $f(x_1, x_2) < B_f$*) and the output of the T2FNN (*u_f*) has an upper bound if the parameter update rules of the T2FNN are chosen as* (7.139)-(7.149) *with the selection of adaptive learning rate (*α*) as in* (7.149)*, then given arbitrary initial conditions* $u_c(0)$*,* $u_c(t)$ *converges asymptotically to zero in a finite time and sliding motion will be achieved.*

The stability conditions in Theorem 7.5 do not need the PD controller to ensure the stability of the control loop and the boundedness of the states of the system any longer. This will greatly relax the conditions required by Theorem 7.4 for the stability of the system. However, note that this theorem is only valid for second-order nonlinear dynamic systems, which is a wide class of nonlinear systems.

7.3.3.1 Proof of Theorem 7.4

As mentioned earlier, the elliptic type-2 MFs are used so that:

$$\underline{\mu}_{1i} = \left(1 - \left|\frac{x_1 - c_{1i}}{d_{1i}}\right|^{a_{2,1i}}\right)^{1/a_{2,1i}} H(x_1, c_{1i}, d_{1i}) \tag{7.151}$$

in which $\underline{\mu}_{1i}$ is the lower bound of the ith MF considered for the error and $H(x_1, c_{1i}, d_{1i})$ is the rectangular function defined as:

$$H(x_1, c, d) = \begin{cases} 1 & c - d < x_1 < c + d \\ 0 & \text{otherwise} \end{cases}$$

If $c_{1i} - d_{1i} < x_1 < c_{1i} + d_{1i}$, then the following equation is valid:

$$ln(\underline{\mu}_{1i}) = \frac{1}{a_{2,1i}} ln\left(1 - \left|\frac{x_1 - c_{1i}}{d_{1i}}\right|^{a_{2,1i}}\right) \tag{7.152}$$

Define $T_{2,1i} = \left|\frac{x_1 - c_{1i}}{d_{1i}}\right|^{a_{2,1i}}$ so that:

$$\dot{T}_{2,1i} = \dot{a}_{2,1i}\left|\frac{x_1 - c_{1i}}{d_{1i}}\right|^{a_{2,1i}} ln\left|\frac{x_1 - c_{1i}}{d_{1i}}\right| + a_{2,1i}\left(\frac{\dot{x}_1 - \dot{c}_{1i}}{|d_{1i}|}\text{sgn}(x_1 - c_{1i}) - \dot{d}_{1i}\frac{|x_1 - c_{1i}|}{d_{1i}^2}\text{sgn}(d_{1i})\right)\left|\frac{x_1 - c_{1i}}{d_{1i}}\right|^{a_{2,1i}-1} \tag{7.153}$$

and the time derivative of $\underline{\mu}_{1i}$ is achieved as:

$$\dot{\underline{\mu}}_{1i} = \frac{-\underline{\mu}_{1i}\dot{a}_{2,1i}}{a_{2,1i}^2} ln\left(1 - \left|\frac{x_1 - c_{1i}}{d_{1i}}\right|^{a_{2,1i}}\right) - \frac{\underline{\mu}_{1i}}{a_{2,1i}}\frac{\dot{T}_{2,1i}}{1 - T_{2,1i}} \tag{7.154}$$

The upper bound of the ith type-2 MFs for e is defined as:

$$\overline{\mu}_{1i} = \left(1 - \left|\frac{x_1 - c_{1i}}{d_{1i}}\right|^{a_{1,1i}}\right)^{1/a_{1,1i}} H(x_1, c_{1i}, d_{1i}) \tag{7.155}$$

Define $T_{1,1i} = \left|\frac{x_1 - c_{1i}}{d_{1i}}\right|^{a_{1,1i}}$ so that:

$$\dot{T}_{1,1i} = \dot{a}_{1,1i}\left|\frac{x_1 - c_{1i}}{d_{1i}}\right|^{a_{1,1i}} ln\left|\frac{x_1 - c_{1i}}{d_{1i}}\right| + a_{1,1i}\left(\frac{\dot{x}_1 - \dot{c}_{1i}}{|d_{1i}|}\text{sgn}(x_1 - c_{1i}) - \dot{d}_{1i}\frac{|x_1 - c_{1i}|}{d_{1i}^2}\text{sgn}(d_{1i})\right)\left|\frac{x_1 - c_{1i}}{d_{1i}}\right|^{a_{1,1i}-1} \tag{7.156}$$

and further:

$$\dot{\overline{\mu}}_{1i} = \frac{-\overline{\mu}_{1i}\dot{a}_{1,1i}}{a_{1,1i}^2} ln\left(1 - \left|\frac{x_1 - c_{1i}}{d_{1i}}\right|^{a_{1,1i}}\right) - \frac{\overline{\mu}_{1i}}{a_{1,1i}}\frac{\dot{T}_{1,1i}}{1 - T_{1,1i}} \tag{7.157}$$

The lower bound of the jth type-2 MFs for $x_2 = \dot{e}$ is defined as:

$$\underline{\mu}_{2j} = \left(1 - \left|\frac{x_2 - c_{2j}}{d_{2j}}\right|^{a_{2,2j}}\right)^{1/a_{2,2j}} H(x_2, c_{2j} - d_{2j}, c_{2j} + d_{2j}) \tag{7.158}$$

Define $T_{2,2j} = \left|\frac{x_2 - c_{2j}}{d_{2j}}\right|^{a_{2,2j}}$ so that:

$$\begin{aligned}\dot{T}_{2,2j} = \dot{a}_{2,2j}\left|\frac{x_2 - c_{2j}}{d_{2j}}\right|^{a_{2,2j}} ln\left|\frac{x_{2j} - c_{2j}}{d_{2j}}\right| + a_{2,2j}\left(\frac{\dot{x}_2 - \dot{c}_{2j}}{|d_{2j}|}\mathrm{sgn}(x_2 - c_{2j})\right. \\ \left. - \dot{d}_{2j}\frac{|x_2 - c_{2j}|}{d_{2j}^2}\mathrm{sgn}(d_{2j})\right)\left|\frac{x_2 - c_{2j}}{d_{2j}}\right|^{a_{2,2j}-1}\end{aligned} \tag{7.159}$$

and further:

$$\dot{\underline{\mu}}_{2j} = \frac{-\underline{\mu}_{2j}\dot{a}_{2,2j}}{a_{2,2j}^2} ln\left(1 - \left|\frac{x_2 - c_{2j}}{d_{2j}}\right|^{a_{2,2j}}\right) - \frac{\underline{\mu}_{2j}}{a_{2,2j}}\frac{\dot{T}_{2,2j}}{1 - T_{2,2j}} \tag{7.160}$$

The upper bound of the jth type-2 MFs for x_2 is defined as:

$$\overline{\mu}_{2j} = \left(1 - \left|\frac{x_2 - c_{2j}}{d_{2j}}\right|^{a_{1,2j}}\right)^{1/a_{1,2j}} H(x_2, c_{2j}, d_{2j}) \tag{7.161}$$

Define $T_{1,2j} = \left|\frac{x_2 - c_{2j}}{d_{2j}}\right|^{a_{1,2j}}$ so that:

$$\begin{aligned}\dot{T}_{1,2j} = \dot{a}_{1,2j}\left|\frac{x_2 - c_{2j}}{d_{2j}}\right|^{a_{1,2j}} ln\left|\frac{x_{2j} - c_{2j}}{d_{2j}}\right| + a_{1,2j}\left(\frac{\dot{x}_2 - \dot{c}_{2j}}{|d_{2j}|}\mathrm{sgn}(x_2 - c_{2j})\right. \\ \left. - \dot{d}_{2j}\frac{|x_2 - c_{2j}|}{d_{2j}^2}\mathrm{sgn}(d_{2j})\right)\left|\frac{x_2 - c_{2j}}{d_{2j}}\right|^{a_{1,2j}-1}\end{aligned} \tag{7.162}$$

and further:

$$\dot{\overline{\mu}}_{2j} = \frac{-\overline{\mu}_{2j}\dot{a}_{1,2j}}{a_{1,2j}^2} ln\left(1 - \left|\frac{x_2 - c_{2j}}{d_{2j}}\right|^{a_{1,2j}}\right) - \frac{\overline{\mu}_{2j}}{a_{1,2j}}\frac{\dot{T}_{1,2j}}{1 - T_{1,2j}} \tag{7.163}$$

We have:

$$\frac{\overline{\mu}_{1i}}{1-T_{1,1i}}\left|\frac{x_1-c_{1i}}{d_{1i}}\right|^{a_{1,1i}-1} = \frac{\overline{\mu}_{1i}-\underline{\mu}_{1i}+\underline{\mu}_{1i}}{1-T_{2,1i}+T_{2,1i}-T_{1,1i}}\left(\left|\frac{x_1-c_{1i}}{d_{1i}}\right|^{a_{2,1i}-1} + \left|\frac{x_1-c_{1i}}{d_{1i}}\right|^{a_{1,1i}-1} - \left|\frac{x_1-c_{1i}}{d_{1i}}\right|^{a_{2,1i}-1}\right)$$
$$= \frac{\underline{\mu}_{1i}}{1-T_{2,1i}}\left|\frac{x_1-c_{1i}}{d_{1i}}\right|^{a_{2,1i}-1}\left(1+\frac{\overline{\mu}_{1i}-\underline{\mu}_{1i}}{\underline{\mu}_{1i}}\right)\times \left(1+\left(\left|\frac{x_1-c_{1i}}{d_{1i}}\right|^{a_{1,1i}-1} - \left|\frac{x_1-c_{1i}}{d_{1i}}\right|^{a_{2,1i}-1}\right)\times \left(\left|\frac{x_1-c_{1i}}{d_{1i}}\right|^{a_{2,1i}-1}\right)^{-1}\right)\left(\frac{1}{1-\frac{T_{1,1i}-T_{2,1i}}{1-T_{2,1i}}}\right) \tag{7.164}$$

Considering the fact that the MFs have nonzero value only if $c_{1i}-d_{1i} < x_1 < c_{1i}+d_{1i}$ we have:

$$\begin{aligned} 0 < T_{1i} < 1 \\ T_{2i} < T_{1i} \end{aligned} \tag{7.165}$$

so that:

$$0 < T_{1i}-T_{2i} < 1-T_{2i} \Rightarrow 0 < \frac{T_{1i}-T_{2i}}{1-T_{2i}} < 1 \tag{7.166}$$

This means it is possible to use the McLauren series as:

$$\frac{\overline{\mu}_{1i}}{1-T_{1,1i}}\left|\frac{x_1-c_{1i}}{d_{1i}}\right|^{a_{1,1i}-1} = \frac{\underline{\mu}_{1i}}{(1-T_{2,1i})}\left|\frac{x_1-c_{1i}}{d_{1i}}\right|^{a_{2,1i}-1}\left(1+\frac{\overline{\mu}_{1i}-\underline{\mu}_{1i}}{\underline{\mu}_{1i}}\right)\times \left(1+\frac{T_{1,1i}-T_{2,1i}}{1-T_{2,1i}}+\left(\frac{T_{1,1i}-T_{2,1i}}{1-T_{2,1i}}\right)^2+H.O.T.\right)\times$$

$$\left(1+\left(\left|\frac{x_1-c_{1i}}{d_{1i}}\right|^{a_{1,1i}-1}-\left|\frac{x_1-c_{1i}}{d_{1i}}\right|^{a_{2,1i}-1}\right)\times\right.$$
$$\left.\left(\left|\frac{x_1-c_{1i}}{d_{1i}}\right|^{a_{2,1i}-1}\right)^{-1}\right) \quad (7.167)$$

It should also be noted that:

$$\begin{aligned}\frac{\overline{\mu}_{1i}}{(1-T_{1,1i})}\left|\frac{x_1-c_{1i}}{d_{1i}}\right|^{a_{1,1i}-1} &= \frac{\underline{\mu}_{1i}}{(1-T_{2,1i})}\left|\frac{x_1-c_{1i}}{d_{1i}}\right|^{a_{2,1i}-1}(1+\Delta_{1i})\\ &= \frac{\underline{\mu}_{1i}}{(1-T_{2,1i})}\left|\frac{x_1-c_{1i}}{d_{1i}}\right|^{a_{2,1i}-1}+\Delta'_{1i} \quad (7.168)\end{aligned}$$

in which:

$$\Delta'_{1i}=\frac{\underline{\mu}_{1i}}{(1-T_{2,1i})}\left|\frac{x_1-c_{1i}}{d_{1i}}\right|^{a_{2,1i}-1}\Delta_{1i} \quad (7.169)$$

and:

$$\begin{aligned}\Delta_{1i} = &\left(1+\frac{\overline{\mu}_{1i}-\underline{\mu}_{1i}}{\underline{\mu}_{1i}}\right)\times\\ &\left(1+\frac{T_{1,1i}-T_{2,1i}}{1-T_{2,1i}}+\left(\frac{T_{1,1i}-T_{2,1i}}{1-T_{2,1i}}\right)^2+H.O.T.\right)\times\\ &\left(1+\left(\left|\frac{x_1-c_{1i}}{d_{1i}}\right|^{a_{1,1i}-1}-\left|\frac{x_1-c_{1i}}{d_{1i}}\right|^{a_{2,1i}-1}\right)\times\right.\\ &\left.\left(\left|\frac{x_1-c_{1i}}{d_{1i}}\right|^{a_{2,1i}-1}\right)^{-1}\right)-1 \quad (7.170)\end{aligned}$$

Furthermore, the following equation should be taken into the account:

$$\begin{aligned}\frac{\overline{\mu}_{2j}}{(1-T_{1,2j})}\left|\frac{x_1-c_{2j}}{d_{2j}}\right|^{a_{1,2j}-1} &= \frac{\underline{\mu}_{2j}}{(1-T_{2,2j})}\left|\frac{x_1-c_{2j}}{d_{2j}}\right|^{a_{2,2j}-1}(1+\Delta_{2j})\\ &= \frac{\underline{\mu}_{2j}}{(1-T_{2,2j})}\left|\frac{x_1-c_{2j}}{d_{2j}}\right|^{a_{2,2j}-1}+\Delta'_{2j}\end{aligned} \quad (7.171)$$

in which:

$$\begin{aligned}\Delta_{2j} &= \left(1+\frac{\overline{\mu}_{2j}-\underline{\mu}_{2j}}{\underline{\mu}_{2j}}\right)\times\\ &\left(1+\frac{T_{1,2j}-T_{2,2j}}{1-T_{2,2j}}+\left(\frac{T_{1,2j}-T_{2,2j}}{1-T_{2,2j}}\right)^2+H.O.T.\right)\times\\ &\left(1+\left(\left|\frac{x_1-c_{2j}}{d_{2j}}\right|^{a_{1,2j}-1}-\left|\frac{x_1-c_{2j}}{d_{2j}}\right|^{a_{2,2j}-1}\right)\times\right.\\ &\left.\left(\left|\frac{x_1-c_{2j}}{d_{2j}}\right|^{a_{2,2j}-1}\right)^{-1}\right)-1 \end{aligned} \tag{7.172}$$

$$\Delta'_{2j}=\frac{\underline{\mu}_{2j}}{(1-T_{2,2j})}\left|\frac{x_1-c_{2j}}{d_{2j}}\right|^{a_{2,2j}-1}\Delta_{2j} \tag{7.173}$$

The following adaptation laws are considered:

$$\dot{a}_{2,1i}=\gamma\left\{\frac{ln\left(1-\left|\frac{x_1-c_{1i}}{d_{1i}}\right|^{a_{2,1i}}\right)}{a_{2,1i}^2}+\frac{\left|\frac{x_1-c_{1i}}{d_{1i}}\right|^{a_{2,1i}}ln\left|\frac{x_1-c_{1i}}{d_{1i}}\right|}{a_{2,1i}(1-T_{2,1i})}\right\}^{-1}H(x_1,c_{1i},d_{1i}) \tag{7.174}$$

$$\dot{a}_{1,1i}=\gamma\left\{\frac{ln\left(1-\left|\frac{x_1-c_{1i}}{d_{1i}}\right|^{a_{1,1i}}\right)}{a_{1,1i}^2}+\frac{\left|\frac{x_1-c_{1i}}{d_{1i}}\right|^{a_{1,1i}}ln\left|\frac{x_1-c_{1i}}{d_{1i}}\right|}{a_{1,1i}(1-T_{1,1i})}\right\}^{-1}H(x_1,c_{1i},d_{1i}) \tag{7.175}$$

$$\dot{a}_{2,2j}=\gamma\left\{\frac{ln\left(1-\left|\frac{x_2-c_{2j}}{d_{2j}}\right|^{a_{2,2j}}\right)}{a_{2,2j}^2}+\frac{\left|\frac{x_2-c_{2j}}{d_{2j}}\right|^{a_{2,2j}}ln\left|\frac{x_2-c_{2j}}{d_{2j}}\right|}{a_{2,2j}(1-T_{2,2j})}\right\}^{-1}H(x_2,c_{2j},d_{2j}) \tag{7.176}$$

$$\dot{a}_{1,2j} = \gamma \left\{ \frac{ln\left(1 - \left|\frac{x_2 - c_{2j}}{d_{2j}}\right|^{a_{1,2j}}\right)}{a_{1,2j}^2} + \frac{\left|\frac{x_2 - c_{2j}}{d_{2j}}\right|^{a_{1,2j}} ln\left|\frac{x_2 - c_{2j}}{d_{2j}}\right|}{a_{1,2j}(1 - T_{1,2j})} \right\}^{-1} H(x_2, c_{2j}, d_{2j}) \tag{7.177}$$

$$\dot{c}_{1i} = -\gamma \frac{|d_{1i}|^{a_{2,1i}}(1 - T_{2,1i})\text{sgn}(x_1 - c_{1i})}{|x_1 - c_{1i}|^{a_{2,1i}-1}} \tag{7.178}$$

$$\dot{d}_{1i} = \gamma \frac{(1 - T_{2,1i})|d_{1i}|^{a_{2,1i}+1}}{|x_1 - c_{1i}|^{a_{2,1i}}} \text{sgn}(d_{1i}) \tag{7.179}$$

$$\dot{c}_{2j} = -\gamma \frac{|d_{2j}|^{a_{2,2j}}(1 - T_{2,2j})\text{sgn}(x_2 - c_{2j})}{|x_2 - c_{2j}|^{a_{2,2j}-1}} \tag{7.180}$$

$$\dot{d}_{2j} = \gamma \frac{(1 - T_{2,2j})|d_{1i}|^{a_{2,2j}+1}}{|x_2 - c_{2j}|^{a_{2,2j}}} \text{sgn}(d_{2j}) \tag{7.181}$$

Considering these adaptation laws we have:

$$\dot{\underline{\mu}}_{1i} = -\gamma \underline{\mu}_{1i} - \frac{\underline{\mu}_{1i}}{1 - T_{2,1i}} \left| \frac{x_1 - c_{1i}}{d_{1i}} \right| \dot{x}_1 \text{sgn}(x_1 - c_{1i}) \tag{7.182}$$

$$\dot{\underline{\mu}}_{2j} = -\gamma \underline{\mu}_{2j} - \frac{\underline{\mu}_{2j}}{1 - T_{2,2j}} \left| \frac{x_2 - c_{2j}}{d_{2j}} \right| \dot{x}_2 \text{sgn}(x_1 - c_{2j}) \tag{7.183}$$

$$\dot{\overline{\mu}}_{1i} = -\gamma \overline{\mu}_{1i} + \Lambda_{1i} \tag{7.184}$$

where Λ_{1i} is defined as:

$$\Lambda_{1i} = \gamma \overline{\mu}_{1i} - \gamma \underline{\mu}_{1i} - \frac{\overline{\mu}_{1i}}{1 - T_{1,1i}} \left| \frac{x_1 - c_{1i}}{d_{1i}} \right| \dot{x}_1 \text{sgn}(x_1 - c_{1i}) \tag{7.185}$$

and further:

$$\dot{\overline{\mu}}_{2j} = -\gamma \underline{\mu}_{2j} - \frac{\overline{\mu}_{2j}}{1 - T_{1,2j}} \left| \frac{x_2 - c_{2j}}{d_{2j}} \right| \dot{x}_2 \text{sgn}(x_2 - c_{2j}) \tag{7.186}$$

so that:

$$\dot{\overline{\mu}}_{2j} = -\gamma\overline{\mu}_{2j} + \Lambda_{2j} \tag{7.187}$$

in which Λ_{2j} is defined as follows:

$$\Lambda_{2j} = \gamma\underline{\mu}_{2j} - \gamma\overline{\mu}_{2j} - \frac{\overline{\mu}_{2j}}{1 - T_{1,2j}}\left|\frac{x_2 - c_{2j}}{d_{2j}}\right|\dot{x}_2\mathrm{sgn}(x_2 - c_{2j}) \tag{7.188}$$

In addition, by definition of $\widetilde{\underline{W}}_{ij}$ and $\widetilde{\overline{W}}_{ij}$ as:

$$\widetilde{\underline{W}}_{ij} = \frac{\underline{W}_{ij}}{\sum_{i=1}^{I}\sum_{j=1}^{J}\underline{W}_{ij}}$$

$$\widetilde{\underline{W}}_{ij} = \frac{\underline{W}_{ij}}{\sum_{i=1}^{I}\sum_{j=1}^{J}\underline{W}_{ij}}$$

the time derivatives of $\widetilde{\underline{W}}_{ij}$ and $\widetilde{\overline{W}}_{ij}$ are obtained as:

$$\widetilde{\underline{W}}_{ij} = \frac{\underline{W}_{ij}}{\sum_{i=1}^{I}\sum_{j=1}^{J}\underline{W}_{ij}} \Rightarrow$$

$$\dot{\widetilde{\underline{W}}}_{ij} = \frac{\left(\underline{\mu}_{1i}(e)\underline{\mu}_{2j}(\dot{e})\right)'\left(\sum_{i=1}^{I}\sum_{j=1}^{J}\underline{W}_{ij}\right)}{\left(\sum_{i=1}^{I}\sum_{j=1}^{J}\underline{W}_{ij}\right)^2} - \frac{\underline{W}_{ij}\left(\sum_{i=1}^{I}\sum_{j=1}^{J}\underline{\mu}_{1i}(e)\underline{\mu}_{2j}(\dot{e})\right)'}{\left(\sum_{i=1}^{I}\sum_{j=1}^{J}\underline{W}_{ij}\right)^2} \tag{7.189}$$

$$\widetilde{\overline{W}}_{ij} = \frac{\overline{W}_{ij}}{\sum_{i=1}^{I}\sum_{j=1}^{J}\overline{W}_{ij}} \Rightarrow$$

$$\dot{\widetilde{\overline{W}}}_{ij} = \frac{\left(\overline{\mu}_{1i}(e)\overline{\mu}_{2j}(\dot{e})\right)'\left(\sum_{i=1}^{I}\sum_{j=1}^{J}\overline{W}_{ij}\right)}{\left(\sum_{i=1}^{I}\sum_{j=1}^{J}\overline{W}_{ij}\right)^2} - \frac{\overline{W}_{ij}\left(\sum_{i=1}^{I}\sum_{j=1}^{J}\overline{\mu}_{1i}(e)\overline{\mu}_{2j}(\dot{e})\right)'}{\left(\sum_{i=1}^{I}\sum_{j=1}^{J}\overline{W}_{ij}\right)^2} \tag{7.190}$$

Since $\tilde{\underline{W}}_{ij} = (\underline{W}_{ij})/(\sum_{i=1}^{I}\sum_{j=1}^{J}\underline{W}_{ij})$ and $\tilde{\overline{W}}_{ij} = (\overline{W}_{ij})/(\sum_{i=1}^{I}\sum_{j=1}^{J}\overline{W}_{ij})$ we have:

$$\dot{\widetilde{\underline{W}}}_{ij} = \frac{\dot{\underline{\mu}}_{1i}(e)\underline{\mu}_{2j}(\dot{e}) + \underline{\mu}_{1i}(e)\dot{\underline{\mu}}_{2j}(\dot{e})}{\sum_{i=1}^{I}\sum_{j=1}^{J}\underline{W}_{ij}} - \frac{\widetilde{\underline{W}}_{ij}\left(\sum_{i=1}^{I}\sum_{j=1}^{J}\left(\dot{\underline{\mu}}_{1i}(e)\underline{\mu}_{2j}(\dot{e}) + \underline{\mu}_{1i}(e)\dot{\underline{\mu}}_{2j}(\dot{e})\right)\right)}{\sum_{i=1}^{I}\sum_{j=1}^{J}\underline{W}_{ij}}$$

$$\dot{\widetilde{\overline{W}}}_{ij} = \frac{\dot{\overline{\mu}}_{1i}(e)\overline{\mu}_{2j}(\dot{e}) + \overline{\mu}_{1i}(e)\dot{\overline{\mu}}_{2j}(\dot{e})}{\sum_{i=1}^{I}\sum_{j=1}^{J}\overline{W}_{ij}} - \frac{\widetilde{\overline{W}}_{ij}\left(\sum_{i=1}^{I}\sum_{j=1}^{J}\left(\dot{\overline{\mu}}_{1i}(e)\overline{\mu}_{2j}(\dot{e}) + \overline{\mu}_{1i}(e)\dot{\overline{\mu}}_{2j}(\dot{e})\right)\right)}{\sum_{i=1}^{I}\sum_{j=1}^{J}\overline{W}_{ij}}$$

The following Lyapunov function is considered to analyze the stability of the system:

$$V_c = \frac{1}{2}u_c^2(t) + \frac{1}{2\gamma_1}(\alpha - \alpha^*)^2 \tag{7.191}$$

The time derivative of the Lyapunov function (7.191) is derived as:

$$\dot{V}_c = u_c\dot{u}_c = u_c(\dot{u}_f + \dot{u}) + \frac{\dot{\alpha}}{\gamma_1}(\alpha - \alpha^*) \tag{7.192}$$

It should be noted that we have:

$$\dot{u}_f = q\sum_{i=1}^{I}\sum_{j=1}^{J}(\dot{f}_{ij}\tilde{\underline{W}}_{ij} + f_{ij}\dot{\tilde{\underline{W}}}_{ij}) + (1-q)\sum_{i=1}^{I}\sum_{j=1}^{J}(\dot{f}_{ij}\widetilde{\overline{W}}_{ij} + f_{ij}\dot{\widetilde{\overline{W}}}_{ij}) \tag{7.193}$$

So the time derivative of the Lyapunov function can be further manipulated as:

$$\dot{V}_c \le u_c(q\sum_{i=1}^{I}\sum_{j=1}^{J}\dot{f}_{ij}\tilde{\underline{W}}_{ij} + (1-q)\sum_{i=1}^{I}\sum_{j=1}^{J}\dot{f}_{ij}\widetilde{\overline{W}}_{ij})$$

$$+ |u_c|(|B_u| + B_f B_{\dot{x}} \frac{B_x + B_c}{B_{d,\min}} + B_f) + \frac{\dot{\alpha}}{\gamma_1}(\alpha - \alpha^*) \tag{7.194}$$

in which:

$$\begin{aligned} |f_{ij}| < B_f, \ |x_1| < B_x, \ |x_2| < B_x, \\ |\dot{x}_1| < B_{\dot{x}}, \ |\dot{x}_2| < B_{\dot{x}}, \\ B_{d,\min} < |d_{1i}|, \ B_{d,\min} < |d_{2j}|, \ |c_{1i}| < B_c, \\ |c_{2j}| < B_c, \ |u| < B_u, \end{aligned} \tag{7.195}$$

In addition, the following adaptation law for f_{ij} is considered:

$$\dot{f}_{ij} = -\frac{q\underline{\widetilde{W}}_{ij} + (1-q)\overline{\widetilde{W}}_{ij}}{\Pi_{ij}^{\mathrm{T}}\Pi_{ij}}\alpha \operatorname{sign}(u_c) \tag{7.196}$$

in which:

$$\Pi_{ij} = \left(\sum_{i=1}^{I}\sum_{j=1}^{J}\left(q\underline{\widetilde{W}}_{ij} + (1-q)\overline{\widetilde{W}}_{ij}\right)\right) \tag{7.197}$$

So that we have the following equation:

$$\dot{V}_c \leq -\alpha|u_c| + B|u_c| + \frac{\dot{\alpha}}{\gamma_1}(\alpha - \alpha^*) \tag{7.198}$$

in which:

$$B = |B_u| + B_f B_{\dot{x}} \frac{B_x + B_c}{B_{d,\min}} + B_f \tag{7.199}$$

It is considered that α^* is as large as:

$$B < \frac{\alpha^*}{2} \tag{7.200}$$

and the adaptation law for the learning rate α is considered to be:

$$\dot{\alpha} = 2\gamma_1|u_c| \tag{7.201}$$

so that:

$$\dot{V}_c \leq -\frac{\alpha}{2}|u_c| \tag{7.202}$$

This means u_c converges asymptotically to zero and this ends the proof.

7.3.3.2 Proof of Theorem 7.5

Consider a second-order system as:

$$\begin{aligned} \dot{x}_1 &= x_2 \\ \dot{x}_2 &= f(x_1, x_2) + u \end{aligned} \tag{7.203}$$

As noted earlier e and $\dot{e}$ are defined as:

$$e = x_d - x_1, \quad \dot{e} = \dot{x}_d - \dot{x}_1 \tag{7.204}$$

so we get the following equation:

$$\ddot{e} = \ddot{x}_d - f(x_1, x_2) - u \tag{7.205}$$

Let the candidate Lyapunov function be in the following form:

$$V = \frac{1}{2}K_p e^2 + \frac{1}{2}\dot{e}^2 + \frac{1}{2\gamma}(K_r - K_r^*)^2, \ \ 0 < \gamma \tag{7.206}$$

The time derivative of this Lyapunov function is calculated as:

$$\dot{V} = K_p e\dot{e} + \dot{e}(\ddot{x}_d - f(x_1, x_2) - u) + \frac{1}{\gamma}\dot{K}_r(K_r - K_r^*) \tag{7.207}$$

The control signal consists of three parts: a classical controller, which is here PD, the output of T2FLS (u_f), and the term to guarantee the robustness (u_r):

$$u = K_p e + K_d \dot{e} + u_r - u_f \tag{7.208}$$

so that:

$$\dot{V} = -K_d \dot{e}^2 + \dot{e}(\ddot{x}_d - f(x_1, x_2) - u_r - u_f) + \frac{1}{\gamma}\dot{K}_r(K_r - K_r^*) \tag{7.209}$$

Here, u_r is added for robust stability and has the following form:

$$u_r = K_r sign(\dot{e}) + \ddot{x}_d \tag{7.210}$$

It is considered that $f(x_1, x_2)$ is bounded in a compact set as $|f(x_1, x_2)| < B_f$:

$$\dot{V} < -K_d\dot{e}^2 + |\dot{e}|(B_f + B_{u_f}) + \ddot{x}_d\dot{e} - u_r\dot{e} + \frac{1}{\gamma}\dot{K}_r(K_r - K_r^*) \tag{7.211}$$

Considering (7.210) we have:

$$\dot{V} < -K_d\dot{e}^2 - K_r|\dot{e}| + K_r^*|\dot{e}| + \frac{1}{\gamma}\dot{K}_r(K_r - K_r^*) \tag{7.212}$$

It is further considered that K_r^* is the final and unknown value of K_r, which is defined as:

$$B_f + B_{u_f} < K_r^* \tag{7.213}$$

The adaptation law for K_r is taken as follows:

$$\dot{K}_r = \gamma|\dot{e}| \tag{7.214}$$

In this way, we have the following equation for the time derivative of the Lyapunov function:

$$\dot{V} < -K_d\dot{e}^2 \tag{7.215}$$

This concludes the stability analysis and means the error signal will converge to zero.

7.4 CONCLUSION

In this chapter, some novel parameter adaptation laws for T2FNN are derived for both identification and control purposes. These adaptation laws benefit from SMC theory. The use of such a nonlinear control approach makes it possible to benefit from its well-established mathematical stability analysis. In the identification part, a novel fully sliding mode parameter update rules are used for the training of an interval T2FNN. The stability analysis of the training is done by using an appropriate Lyapunov function. In the controller design part, in the case when the system is of *n*th order, the stability of the training of the T2FNN is considered. Furthermore, the stability of the system (not only the stability of the learning but also the overall stability of the system), in the case of having a second-order system, is shown.

REFERENCES

[1] E. Kayacan, O. Cigdem, O. Kaynak, Sliding mode control approach for online learning as applied to type-2 fuzzy neural networks and its experimental evaluation, IEEE Trans. Indust. Electron. 59 (9) (2012) 3510-3520.

[2] O. Kaynak, K. Erbatur, M. Ertugnrl, The fusion of computationally intelligent methodologies and sliding-mode control—a survey, IEEE Trans. Indust. Electron. 48 (1) (2001) 4-17.

[3] X. Yu, O. Kaynak, Sliding-mode control with soft computing: a survey, IEEE Trans. Indust. Electron. 56 (9) (2009) 3275-3285.

[4] S. Masumpoor, H. yaghobi, M.A. Khanesar, Adaptive sliding-mode type-2 neuro-fuzzy control of an induction motor, Expert Syst. Appl. ISSN 0957-4174.

[5] M. Begian, W. Melek, J. Mendel, Parametric design of stable type-2 TSK fuzzy systems, in: Fuzzy Information Processing Society, 2008. NAFIPS 2008. Annual Meeting of the North American, 2008, pp. 1-6.

[6] E. Kayacan, M. Khanesar, Identification of nonlinear dynamic systems using type-2 fuzzy neural networks a novel learning algorithm and a comparative study, IEEE Trans. Indust. Electron. 62 (3) (2015) 1716-1724.

[7] J.-J. Slotine, W. Li, Applied Nonlinear Control, Prentice Hall, Upper Saddle River, NJ, 1991.

[8] M. Biglarbegian, W. Melek, J. Mendel, On the stability of interval type-2 tsk fuzzy logic control systems, IEEE Trans. Syst. Man Cybernet. B Cybernet. 40 (3) (2010) 798-818.

[9] V. Utkin, Sliding Modes in Control Optimization, Springer-Verlag, New York, 1992.

CHAPTER 8

Hybrid Training Method for Type-2 Fuzzy Neural Networks Using Particle Swarm Optimization

Contents

Abstract

Even if we have highly capable computers today, nature still has the best engineering designs. For example, it is not coincidence that the nose of an airplane is very similar to that of a dolphin. Not only the physical appearance but also the behavior of animals

Fuzzy Neural Networks for Real Time Control Applications
http://dx.doi.org/10.1016/B978-0-12-802687-8.00008-6

in nature are opening the doors for new theories. This chapter shows that nature is still helping humans make the most efficient and brilliant engineering designs.

In this chapter a novel hybrid training method based on continuous version of PSO and SMC-based training method for T2FNNs is proposed. The proposed approach uses PSO for the training of the antecedent parts of T2FNNs, which appear nonlinearly in the output of T2FNNs and SMC-based training methods for the training of parameters of their consequent parts. The use of PSO makes it possible to lessen the probability of entrapment of the parameters in local minima while proposing simple adaptation laws for the parameters of the antecedent parts of T2FNN as compared to popular approaches like GD, LM and extended KF. Moreover, the SMC theory-based rules for the parameters of the consequent part are simple, have closed-form and benefit from a rigorous mathematics. In addition, using SMC theory-based, the responsible parameter for sharing the contributions of the lower and upper parts of the type-2 fuzzy MFs is tuned as well. The stability of the proposed hybrid training method is proved by using an appropriate Lyapunov function.

Keywords

Hybrid training, Particle swarm optimization, Type-2 fuzzy neural networks, Levenberge-Marquardt

8.1 INTRODUCTION

Even if we have highly capable computers today, nature still has the best engineering designs. For example, it is not coincidence that the nose of an airplane is very similar to that of a dolphin. Not only the physical appearance but also the behavior of animals in nature are opening the doors for new theories. This chapter shows that nature is still helping humans make the most efficient and brilliant engineering designs.

The PSO is a population-based search algorithm. This optimization algorithm is based on the observations made from the flight of birds within a flock, school of fish, and so on. Each individual bird in a flock flies independently and unpredictably, but in such a way that each stays connected and forms a social group. What can be observed from birds' behavior when flying or fish when swimming is that they have tendency to move to the best position for their neighbors. They even communicate with each other in order to obtain the best position and continue flying synchronously.

PSO has become one of the most popular and most preferable optimization algorithms. Since it searches a large space of possible solutions, its applications to most real-world applications have been successful, and it has outperformed other meta-heuristic optimization algorithms considering the precision of the solutions and speed of convergence.

The simplified version of this concept is formulated with a two-state dynamic system. The two state dynamic system is comprised of the position as the first state and velocity as the second state of the system. It is then assumed that each particle has a multidimensional position vector. The number of dimensions considered for each particle is the same as the dimensions of the optimization problem. The optimality is also defined based on the objective function to be optimized. PSO requires no assumptions, or just a few assumptions, about the objective function to be optimized. The change in the position of a particle is defined by its corresponding velocity vector, which determines the moving direction of the particle.

The moving direction has three different terms that can be put into two categories: exploration and exploitation terms. The exploration is basically composed of a momentum term that tries to maintain the latest moving direction of the particle which is a random direction and makes it possible to visit more points. The exploitation terms are the terms that direct a particle toward its own best experience visited and toward the best position found by all particles in the swarm. Since the best position found in the first epochs are far from the true optimum of the objective function, the gain of exploration term is higher than that of exploitation to visit more random positions at first few epochs. However, as more points are visited by the swarm the best experience of each individual and the best experience of the whole swarm are more likely to be meaningful. Hence, at larger values of epochs the exploitation terms become dominant.

The original version of PSO was discrete time. However, its continuous version has also been studied in number of papers, making it easier to analyze mathematically. Furthermore, the stability analysis of its discrete and continuous version has been studied. It has been shown that in order to have stable behavior for each particle toward best position found by individuals and the best position found by all members of a swarm, certain constrains must be considered in the selection of the parameters of PSO.

One of the applications that PSO could be successfully used for is the estimation of the parameters of FNNs in which PSO is used alone or in combination with other estimation algorithms.

8.1.1 Fully PSO Training Algorithms

In these methods PSO is used to estimate all the parameters of FNNs. They are easy to implement because of the simple nature of the PSO method and do not necessitate the gradient calculation of FNN output with respect to its parameters as is common with most computation-based training algorithms. Moreover, since PSO proposes multiple evolving individuals as the solution to the problem, the solutions entrap less in local minima. However, multiple solutions require more memory to be used. Moreover, they require multiple feed-forward calculation of FNN, which is time consuming. Examples of PSO-based methods for training FNNs can be found in Refs [1–4].

8.1.2 Hybrid PSO with Computation-based Training Methods

These methods benefit from the advantages of the above mentioned algorithm. In these approaches, the population-based algorithms are used for the estimation of the antecedent part parameters because they are gradient-free. In other words, since the parameters of the antecedent part appear nonlinearly in the output of FNN, the calculation of the gradient of the cost function with respect to them is difficult and computation-based methods may even fail to optimize these parameters globally. However, since the parameters of the consequent part appear linearly in the output of a FNN, the computation-based method can be easily applied to them. The use of these methods for the consequent part parameters makes it possible to benefit from the mathematical background of these methods. Moreover, the use of computational methods for the estimation of the parameters of the consequent part instead of using population-based methods highly reduces the dimension of individuals. Examples of hybrid training methods using PSO with other computation-based approaches that are applied to the training of FNNs can be found in Refs [5–7].

8.2 CONTINUOUS VERSION OF PARTICLE SWARM OPTIMIZATION

In this section, the continuous version of PSO is briefly discussed, and its mathematical formulation is investigated. The dynamics of each dimension of particles in this optimization algorithm are independent of each other. The position of the lth dimension of the sth particle is denoted by x_{sl} and its velocity is depicted using v_{sl}. Therefore, the dynamics of the lth dimension of the sth particle are formulated as follows [8]:

$$\begin{aligned}
\dot{x}_{sl}(t) &= v_{sl}(t) \\
\dot{v}_{sl}(t) &= (\beta - 1)v_{sl}(t) + \gamma_1 \left(Pl_{sl}(t) - x_{sl}\right) + \gamma_2 \left(Pg_{sl}(t) - x_{sl}\right) \\
x_{sl}(0) &= x_{sl0}, v_{sl}(0) = v_{sl0}, \ l = 1, \ldots, n, \ s = 1, \ldots, N
\end{aligned} \tag{8.1}$$

in which $Pl_{sl}(t)$ is the lth dimension of the best particle achieved for sth particle and $Pg_{sl}(t)$ is the lth dimension of the best particle achieved for all particles. Furthermore, $Pl_s(t) = [Pl_{s1}, Pl_{s2}, \ldots, Pl_{sn}]$, $Pg(t) = [Pg_1, Pg_2, \ldots, Pg_n]$, and n is the number of dimensions. $Pl_s(t)$ and $Pg(t)$ are updated as follows:

$$
\begin{aligned}
Pl_s(t) &= \arg\min\{f(Pl_s(t^-)), f(x_s(t))\} \\
Pg(t) &= \arg\min\{f(Pg(t^-)), f(x_1(t)), \ldots, f(x_N(t))\}
\end{aligned}
$$

in which N is the number of individuals in the swarm, $f(.)$ is the cost function to be minimized. If $P_{sl} = \gamma_1 Pl_{sl}(t) + \gamma_2 Pg_{sl}(t)$ and $\alpha = \gamma_1 + \gamma_2$, the mathematical description of continuous time PSO is modified as follows:

$$
\begin{aligned}
\dot{x}_{sl}(t) &= v_{sl}(t) \\
\dot{v}_{sl}(t) &= (\beta - 1)v_{sl}(t) - \alpha x_{sl} + \alpha P_{sl}(t) \\
x_{sl}(0) &= x_{sl0}, v_{sl}(0) = v_{sl0}, \ l = 1, \ldots, n, \ s = 1, \ldots, N
\end{aligned} \tag{8.2}
$$

8.3 ANALYSIS OF CONTINUOUS VERSION OF PARTICLE SWARM OPTIMIZATION

8.3.1 Stability Analysis

In order to analyze the stability of CPSO, the equation of motion of PSO (8.2) can be reformulated as follows:

$$
\ddot{x}_{sl}(t) - (\beta - 1)\dot{x}_{sl}(t) + \alpha x_{sl} = \alpha P_{sl}(t) \tag{8.3}
$$

which is a second-order dynamic system whose input is $P_{sl}(t)$. Since $P_{sl}(t)$ is selected from the best visited positions met by an individual member of the swarm and by the whole swarm, it is finite if its corresponding position vector $x_{sl}(t)$ is finite and stable. On the other hand, (8.3) shows that the governing equations of the position vector $x_{sl}(t)$ is a second-order dynamic system. Using the Routh-Hurwitz stability analysis, in order to have a stable motion, the coefficients of $x_{sl}(t)$ and its derivatives must be positive. Therefore, the stability conditions are as follows:

$$
\begin{aligned}
\beta &< 1 \\
\alpha &> 0
\end{aligned} \tag{8.4}
$$

8.3.2 Dynamic Behavior

Since the basic CPSO can be considered as a linear second-order system, the damping ratio of this system is equal to $\frac{1-\beta}{2\sqrt{\alpha}}$ and its natural frequency is obtained as $\sqrt{\alpha}$. If $0 < \frac{1-\beta}{2\sqrt{\alpha}} < 1$, the motion of a single dimension of a particle oscillates around P_{sl}, and finally converges toward it in infinite time. This motion behavior is called under-damped motion behavior. However, if $\frac{1-\beta}{2\sqrt{\alpha}} > 1$ the system converges slowly, but without any oscillations toward $P_{sl}(t)$. This behavior is called over-damped. Figure 8.1 presents the motion behavior of one dimension of a single particle with respect to time and its phase plane when it is drawn with respect to its corresponding velocity. The P_{sl} is taken as $P_{sl} = 1$; in other words, the reference signal for the lth dimension of the position of the particle x_{sl} is taken equal to 1. As can be seen from Fig. 8.1, different values for the β and α parameters may result in different motion behavior in x_{sl}. The motion behavior may be without any oscillation but slow or over-damped (Fig. 8.1(a), (b)), fast but oscillating (Fig. 8.1(c), (d)) or super-fast but with high level of oscillations (Fig. 8.1(e), (f)). The two latter behaviors are called under-damped. As expected, the continuous version of PSO acts differently when different values for poles are used.

8.3.3 Higher-Order Continuous PSO

In order to increase the degrees of freedom of basic CPSO, HCPSO can be used in which the dynamics of x_{ls} are modified as follows:

$$\begin{aligned} \dot{x}_{sl} &= \nu_{sl} \\ \dot{\nu}_{sl} &= (\beta - 1)\nu_{sl} - \alpha x_{sl} + \alpha P - \alpha_2 \nu_{sl}^3 \end{aligned} \tag{8.5}$$

Type the existence of the term ν_{sl}^3 makes it possible for HCPSO to cover a wider range of behaviors for a particle. Moreover, the existence of this term makes it possible to have a wider selection of possibilities for the parameters of PSO. For example, Fig. 8.2 shows the behavior of basic CPSO and HCPSO when the value of $\beta = 1.1$. As can be seen from this figure, although the response of CPSO is unstable with this selection for w, no instability happens in HCPSO. Another benefit of using HCPSO over CPSO is that it is possible to achieve a limit cycle that is not possible in a linear system such as CPSO. In this way, HCPSO can search a wider area.

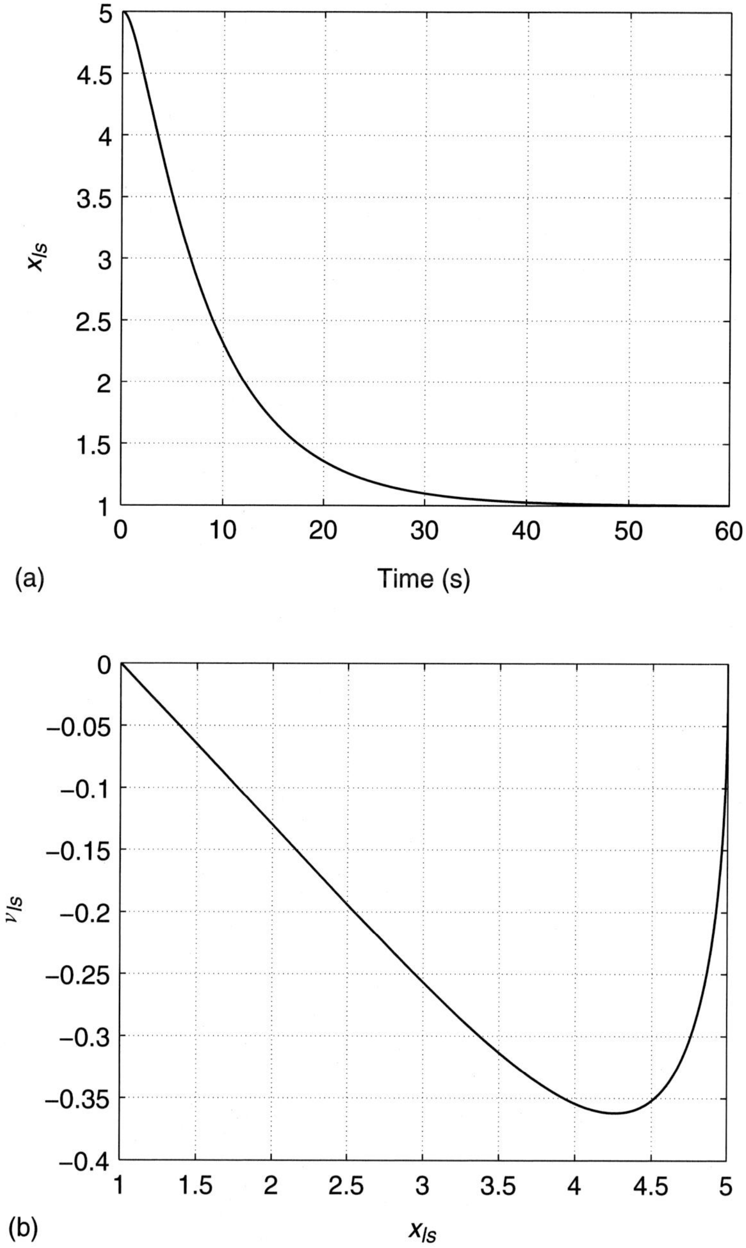

Figure 8.1 (a) Motion behavior of x_{ls} when $\beta = 0.1$ and $\alpha = 0.1$; (b) phase plane output of ν_{ls} with respect to x_{ls} when $\beta = 0.1$ and $\alpha = 0.2$; *(Continued)*

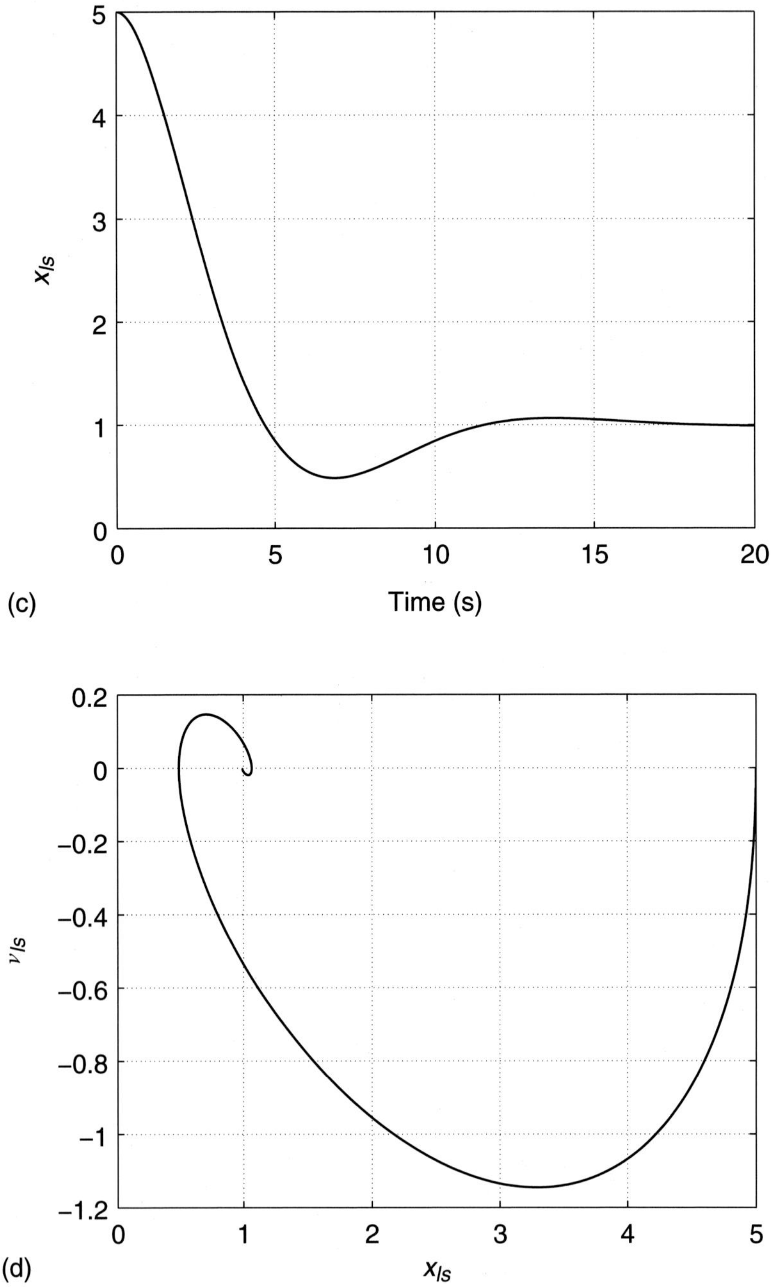

Figure 8.1, cont'd (c) motion behavior of x_{ls} when $\beta = 0.4$ and $\alpha = 0.3$; (d) phase plane output of ν_{ls} with respect to x_{ls} when $\beta = 0.4$ and $\alpha = 0.3$;

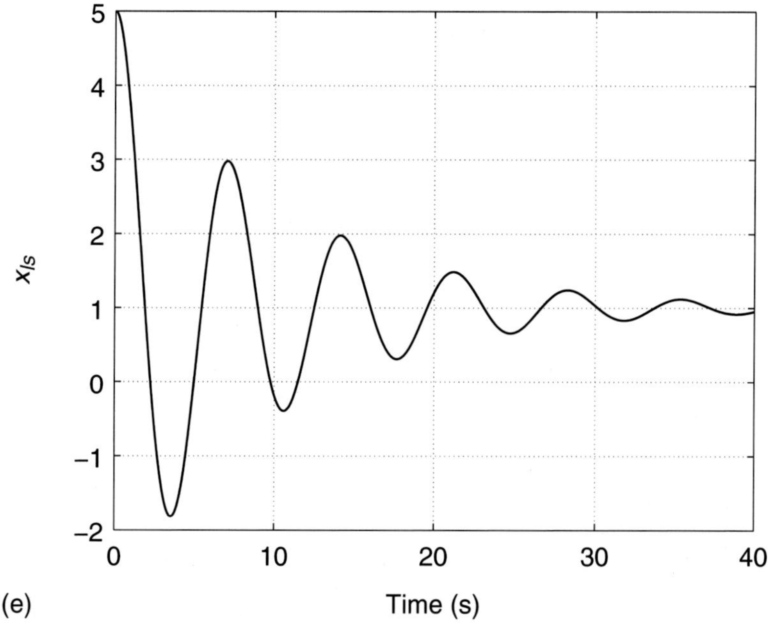

(e)

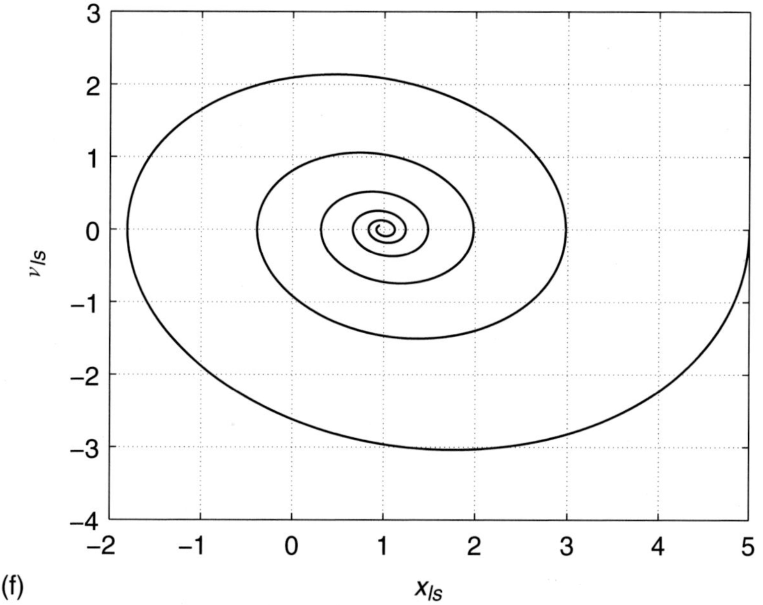

(f)

Figure 8.1, cont'd (e) motion behavior of x_{ls} when $\beta = 0.8$ and $\alpha = 0.8$; (f) phase plane output of ν_{ls} with respect to x_{ls} when $\beta = 0.8$ and $\alpha = 0.8$.

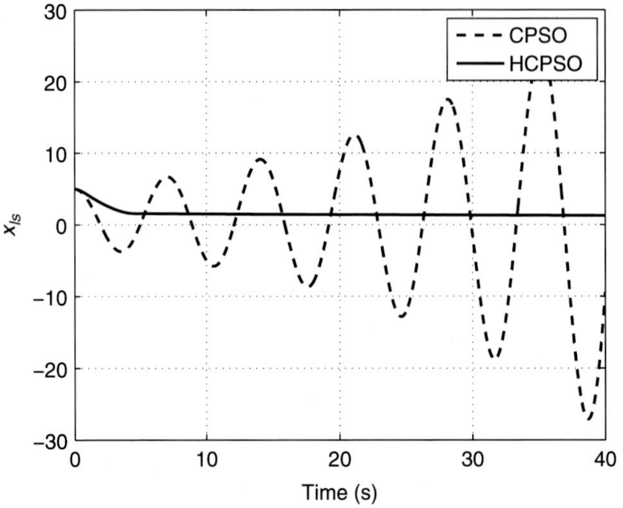

Figure 8.2 Motion behavior of x_{sl} for CPSO and HCPSO.

8.4 PROPOSED HYBRID TRAINING ALGORITHM FOR TYPE-2 FUZZY NEURAL NETWORK

8.4.1 Proposed Hybrid Training Algorithm for T2FNN with Type-2 MFs with Uncertain σ

8.4.1.1 T2FNN with Type-2 MFs with Uncertain σ

The upper and lower type-2 fuzzy Gaussian MFs with an uncertain standard deviation (Fig. 8.3) can be represented as follows:

$$\overline{\mu}_{ik}(x_i) = \exp\left(-\frac{1}{2}\frac{(x_i - c_{ik})^2}{\overline{\sigma}_{ik}^2}\right) \tag{8.6}$$

$$\underline{\mu}_{ik}(x_i) = \exp\left(-\frac{1}{2}\frac{(x_i - c_{ik})^2}{\underline{\sigma}_{ik}^2}\right) \tag{8.7}$$

where c_{ik} is the center value of the *k*th type-2 fuzzy set for the *i*th input. The parameters $\overline{\sigma}_{ik}$ and $\underline{\sigma}_{ik}$ are standard deviations for the upper and lower MFs.

After computing the lower and upper membership degrees $\underline{\mu}$ and $\overline{\mu}$ for each input, the firing strengths of the rules using the *prod* t-norm operator are calculated as follows:

$$\underline{w}_r = \underline{\mu}_{\tilde{A}1}(x_1) * \underline{\mu}_{\tilde{A}2}(x_2) * \cdots \underline{\mu}_{\tilde{A}I}(x_I)$$

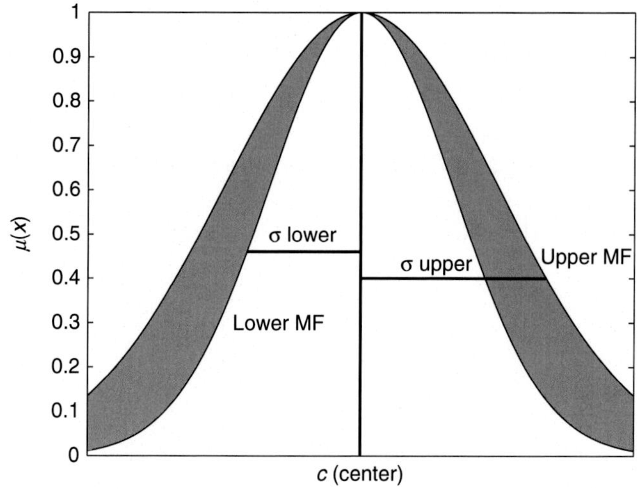

Figure 8.3 Type-2 Gaussian fuzzy MF with uncertain center.

$$\overline{w}_r = \overline{\mu}_{\tilde{A}1}(x_1) * \overline{\mu}_{\tilde{A}2}(x_2) * \cdots \overline{\mu}_{\tilde{A}I}(x_I) \tag{8.8}$$

The consequent part corresponding to each fuzzy rule is a linear combination of the inputs $x_1, x_2, \ldots, x_I$. This linear function is called f_r and is defined as $f_r = b_r + \sum_{i=1}^{I} a_{ri}x_i$. The output of the network is approximated as follows:

$$y_N = q\sum_{r=1}^{R} f_r\underline{\tilde{w}}_r + (1-q)\sum_{r=1}^{R} f_r\tilde{\overline{w}}_r \tag{8.9}$$

where R is the number of the rules of T2FNN, and $\underline{\tilde{w}_r}$ and $\tilde{\overline{w}}_r$ are the normalized values of the lower and upper output signals from the second hidden layer of the network as follows:

$$\underline{\tilde{w}_r} = \frac{\underline{w_r}}{\sum_{i=1}^{R} \underline{w_r}} \quad \text{and} \quad \tilde{\overline{w}}_r = \frac{\overline{w_r}}{\sum_{i=1}^{R} \overline{w_r}} \tag{8.10}$$

The following vectors can be specified:
$\underline{\widetilde{W}}(t) = \left[\underline{\widetilde{w_1}}(t)\ \underline{\widetilde{w_2}}(t) \ldots \underline{\widetilde{w_R}}(t)\right]^{\mathrm{T}}$, $\widetilde{\overline{W}}(t) = \left[\widetilde{\overline{w_1}}(t)\ \widetilde{\overline{w_2}}(t) \ldots \widetilde{\overline{w_R}}(t)\right]^{\mathrm{T}}$, and $F = [f_1\, f_2\, \ldots f_R]$

The following assumptions have been used in this investigation: The time derivative of both the input signals and target signal can be considered bounded:

$$|\dot{x}_i(t)| \leq B_{\dot{x}}, |x_i(t)| \leq B_x, \ (i = 1 \dots I) \text{ and } |\dot{y}(t)| \leq B_{\dot{y}} \quad \forall t \tag{8.11}$$

where $B_{\dot{x}}$, B_x, and $B_{\dot{y}}$ are assumed to be some known positive constants.

8.4.1.2 Proposed Hybrid Training Algorithm for T2FNN by Using Type-2 MFs with Uncertain σ

The proposed hybrid PSO and SMC training method is as follows:

$$\begin{aligned} \dot{\underline{\sigma}}_{sl} &= \nu_{\underline{\sigma}_{sl}} \\ \dot{\nu}_{\underline{\sigma}_{sl}} &= (\beta - 1)\nu_{\underline{\sigma}_{sl}} - \alpha\underline{\sigma}_{sl} + \gamma_1 Pl_{\underline{\sigma}_{sl}} + \gamma_2 Pg_{\underline{\sigma}_l} - \alpha_2 \nu^3_{\underline{\sigma}_{sl}}, \quad l = 1, \dots, n, \\ & s = 1, \dots, N \end{aligned} \tag{8.12}$$

$$\begin{aligned} \dot{\overline{\sigma}}_{sl} &= \nu_{\overline{\sigma}_{sl}} \\ \dot{\nu}_{\overline{\sigma}_{sl}} &= (\beta - 1)\nu_{\overline{\sigma}_{sl}} - \alpha\overline{\sigma}_{sl} + \gamma_1 Pl_{\overline{\sigma}_{sl}} + \gamma_2 Pg_{\overline{\sigma}_{sl}} - \alpha_2 \nu^3_{\overline{\sigma}_{sl}}, \quad l = 1, \dots, n, \\ & s = 1, \dots, N \end{aligned} \tag{8.13}$$

$$\begin{aligned} \dot{c}_{sl} &= \nu_{c_{sl}} \\ \dot{\nu}_{c_{sl}} &= (\beta - 1)\nu_{c_{sl}} - \alpha c_{sl} + \gamma_1 Pl_{c_{sl}} + \gamma_2 Pg_{c_l} - \alpha_2 \nu^3_{c_{sl}}, \quad l = 1, \dots, n, \\ & s = 1, \dots, N \end{aligned} \tag{8.14}$$

$$\begin{aligned} \dot{a}_{ri_s} &= -x_i \frac{q\underline{\tilde{w}}_r + (1-q)\overline{\tilde{w}}_r}{\left(q\underline{\tilde{w}}_r + (1-q)\overline{\tilde{w}}_r\right)^{\mathrm{T}}\left(q\underline{\tilde{w}}_r + (1-q)\overline{\tilde{w}}_r\right)} \alpha \operatorname{sgn}(e), \quad s = 1, \dots, N, \\ & r = 1, \dots, R \end{aligned} \tag{8.15}$$

$$\begin{aligned} \dot{b}_{r_s} &= -\frac{q_s\underline{\tilde{w}}_{r_s} + (1-q_s)\overline{\tilde{w}}_{r_s}}{\left(q_s\underline{\tilde{w}}_{r_s} + (1-q_s)\overline{\tilde{w}}_{r_s}\right)^{\mathrm{T}}\left(q_s\underline{\tilde{w}}_{r_s} + (1-q_s)\overline{\tilde{w}}_{r_s}\right)} \left(\alpha_1 e_s + \alpha \operatorname{sgn}(e_s)\right) \\ & s = 1, \dots, N, r = 1, \dots, R \end{aligned} \tag{8.16}$$

$$\dot{q}_s = -\frac{1}{F_s(\underline{\widetilde{W}}_s - \overline{\widetilde{W}}_s)^{\mathrm{T}}} \alpha \operatorname{sgn}(e_s), \quad s = 1, \dots, N \tag{8.17}$$

In (8.12)-(8.17), n is the number of the type-2 MFs, e_s is identification error corresponding to the sth particle, subscript s refers to the sth particle, and subscript l refers to lth dimension. For example, c_{sl} means the lth dimension of the sth particle corresponding to c, the centers of MFs. The cost function $f(e_s(t))$ is defined as:

$$f(e_s(t)) = \frac{1}{2}\sum e_s^2(t), \quad s = 1, \ldots, N \tag{8.18}$$

where the limits of the summation depend on the window size decided by the programmer and the local best experiences of each particle $Pl_{\overline{\sigma}_s}$, $Pl_{\underline{\sigma}_s}$, and Pl_{c_s} are updated using this cost function ($f(.)$) as follows:

$$\begin{aligned} Pl_{\overline{\sigma}_s}, Pl_{\underline{\sigma}_s}, \text{ and } Pl_{c_s} = \arg\min\{& f(e(t)|Pl_{\overline{\sigma}_s}(t^-), Pl_{\underline{\sigma}_s}(t^-), Pl_{c_s}(t^-)), \\ & f(e(t)|\overline{\sigma}_s(t), \underline{\sigma}_s(t), c_s(t))\} \end{aligned} \tag{8.19}$$

$$\begin{aligned} Pg_{\overline{\sigma}}, Pg_{\underline{\sigma}}, \text{ and } Pg_c = \arg\ \min\{& f(e(t)|Pg_{\overline{\sigma}}(t^-), Pg_{\underline{\sigma}}(t^-), Pg_c(t^-)), \\ & f(e(t)|\overline{\sigma}_s(t), \underline{\sigma}_s(t), c_s(t))\} \quad \forall s \in N \end{aligned} \tag{8.20}$$

Theorem 8.1. *The adaptation laws of* (8.12)-(8.17) *guarantee that the parameters of T2FNN in the form of* (8.9) *remain bounded provided that:*

$$|\dot{x}_i(t)| \leq B_{\dot{x}}, |x_i(t)| \leq B_x, \ (i = 1 \ldots I) \text{ and } |\dot{y}(t)| \leq B_{\dot{y}} \quad \forall t \tag{8.21}$$

where $B_{\dot{x}}$, B_x, *and* $B_{\dot{y}}$ *are assumed to be some known positive constants.*

8.4.1.3 Stability Analysis

Since the stability of each particle is independent of the others, the stability of one particle is considered. The time derivative of (8.10) is calculated as follows:

$$\dot{\tilde{\underline{w}}}_r = -\tilde{\underline{w}}_r\underline{K}_r + \tilde{\underline{w}}_r\sum_{r=1}^{R}\tilde{\underline{w}}_r\underline{K}_r; \quad \dot{\tilde{\overline{w}}}_r = -\tilde{\overline{w}}_r\overline{K}_r + \tilde{\overline{w}}_r\sum_{r=1}^{R}\tilde{\overline{w}}_r\overline{K}_r \tag{8.22}$$

where:

$$\underline{A}_{ik} = \frac{x_i - c_{ik}}{\underline{\sigma}_{ik}} \text{ and } \overline{A}_{ik} = \frac{x_i - c_{ik}}{\overline{\sigma}_{ik}}$$

$$\underline{K}_r = \sum_{i=1}^{I} \underline{A}_{ik}\dot{\underline{A}}_{ik} \text{ and } \overline{K}_r = \sum_{i=1}^{I} \overline{A}_{ik}\dot{\overline{A}}_{ik}$$

By using the following Lyapunov function, the stability of a single particle is analyzed:

$$V = \frac{1}{2}e^2 + \underbrace{\frac{\alpha_1}{2}\underline{\sigma}^{\mathrm{T}}\underline{\sigma} + \frac{1}{2}\nu_{\underline{\sigma}}^{\mathrm{T}}\nu_{\underline{\sigma}}}_{I_1} + \underbrace{\frac{\alpha_1}{2}\overline{\sigma}^{\mathrm{T}}\overline{\sigma} + \frac{1}{2}\nu_{\overline{\sigma}}^{\mathrm{T}}\nu_{\overline{\sigma}}}_{I_2} + \underbrace{\frac{\alpha_1}{2}c^{\mathrm{T}}c + \frac{1}{2}\nu_c^{\mathrm{T}}\nu_c}_{I_3} \tag{8.23}$$

where $\overline{\sigma} = [\overline{\sigma}_{11}, \ldots, \overline{\sigma}_{ik}, \ldots, \overline{\sigma}_{IK}]$, $\underline{\sigma} = [\underline{\sigma}_{11}, \ldots, \underline{\sigma}_{ik}, \ldots, \underline{\sigma}_{IK}]$, $c = [c_{11}, \ldots, \overline{c}_{ik}, \ldots, c_{IK}]$, and $I \times K = n$. In order to make the time derivative of the Lyapunov function easier to calculate, the following assumption is made:

$$I_1 = \frac{\alpha_1}{2}\underline{\sigma}^{\mathrm{T}}\underline{\sigma} + \frac{1}{2}\nu_{\underline{\sigma}}^{\mathrm{T}}\nu_{\underline{\sigma}} \tag{8.24}$$

Therefore, we have the following equation for the time derivative of I_1:

$$\dot{I}_1 = \alpha_1 \sum_i \sum_k \underline{\sigma}_{ik}\nu_{\underline{\sigma}_{ik}} + \sum_i \sum_k \nu_{\underline{\sigma}_{ik}}\left((\beta - 1)\nu_{\underline{\sigma}_{ik}} - \alpha_1\underline{\sigma}_{ik} + \alpha P_{\underline{\sigma}_{ik}} - \alpha_2\nu_{\underline{\sigma}_{ik}}^3\right) \tag{8.25}$$

and further:

$$\dot{I}_1 = \sum_i \sum_k \left((\beta - 1)\nu_{\underline{\sigma}_{ik}}^2 + \alpha P_{\underline{\sigma}_{ik}} - \alpha_2\nu_{\underline{\sigma}_{ik}}^4\right) \tag{8.26}$$

Similar to the calculation of $\dot{I}_1$, the following equations for $\dot{I}_2$ and $\dot{I}_3$ are obtained:

$$\dot{I}_2 = \sum_i \sum_k \left((\beta - 1)\nu_{\overline{\sigma}_{ik}}^2 + \alpha P_{\overline{\sigma}_{ik}} - \alpha_2\nu_{\overline{\sigma}_{ik}}^4\right) \tag{8.27}$$

$$\dot{I}_3 = \sum_i \sum_k \left((\beta - 1)\nu_{c_{ik}}^2 + \alpha P_{c_{ik}} - \alpha_2\nu_{c_{ik}}^4\right) \tag{8.28}$$

The time derivative of (8.23) can be calculated as follows:

$$\dot{V} = \dot{e}e + \dot{I}_1 + \dot{I}_2 + \dot{I}_3 = e(\dot{y}_N - \dot{y}) + \dot{I}_1 + \dot{I}_2 + \dot{I}_3 \tag{8.29}$$

Taking the time derivative of (8.9), the following term can be obtained:

$$\begin{aligned}\dot{y}_N =& \dot{q}\sum_{r=1}^{R} f_r\underline{\tilde{w}}_r + q\sum_{r=1}^{R}(\dot{f}_r\underline{\tilde{w}}_r + f_r\dot{\underline{\tilde{w}}}_r) - \dot{q}\sum_{r=1}^{R} f_r\tilde{\overline{w}}_r \\ &+ (1-q)\sum_{r=1}^{R}(\dot{f}_r\tilde{\overline{w}}_r + f_r\dot{\tilde{\overline{w}}}_r)\end{aligned} \tag{8.30}$$

By using (8.22) and (8.30), the following term can be obtained:

$$\begin{aligned}\dot{y}_N =& \dot{q}\sum_{r=1}^{R} f_r\underline{\tilde{w}}_r - \dot{q}\sum_{r=1}^{R} f_r\tilde{\overline{w}}_r + q\sum_{r=1}^{R}\left(\dot{f}_r\underline{\tilde{w}}_r + f_r(-\underline{\tilde{w}}_r\underline{K}_r + \underline{\tilde{w}}_r\sum_{r=1}^{R}\underline{\tilde{w}}_r\underline{K}_r)\right) \\ &+ (1-q)\sum_{r=1}^{R}\left(\dot{f}_r\tilde{\overline{w}}_r + f_r(-\tilde{\overline{w}}_r\overline{K}_r + \tilde{\overline{w}}_r\sum_{r=1}^{R}\tilde{\overline{w}}_r\overline{K}_r)\right) \\ \le& \dot{q}\sum_{r=1}^{R}\left(f_r(\underline{\tilde{w}}_r - \tilde{\overline{w}}_r)\right) + \sum_{r=1}^{R}\left(\dot{f}_r(q\underline{\tilde{w}}_r + (1-q)\tilde{\overline{w}}_r)\right) \\ &+ M_1B_f\sum_{i=1}^{R}\|\underline{K}_r\| + M_2B_f\sum_{i=1}^{R}\|\overline{K}_r\|\end{aligned} \tag{8.31}$$

in which $\|.\|$ denotes the Euclidean norm of its corresponding argument and B_f is such that $\|F\| < B_f$ and the value of $\|F\|$ is kept bounded using a projection algorithm. Considering the fact that x_i, $\dot{x}_i$, $\dot{y}$ are bounded we have:

$$\sum_{i=1}^{R}\|\underline{K}_r\| \le M_3 + M_4\|\dot{c}\| + M_5\|\dot{\underline{\sigma}}\|$$

$$\sum_{i=1}^{R}\|\overline{K}_r\| \le M_6 + M_7\|\dot{c}\| + M_8\|\dot{\overline{\sigma}}\|$$

in which M_3, M_4, M_5, M_6, M_7, and M_8 are positive real values. Considering the adaptation laws as:

$$\dot{a}_{ri} = -x_i\frac{q\underline{\tilde{w}}_r + (1-q)\tilde{\overline{w}}_r}{\left(q\underline{\tilde{w}}_r + (1-q)\tilde{\overline{w}}_r\right)^{\mathrm{T}}\left(q\underline{\tilde{w}}_r + (1-q)\tilde{\overline{w}}_r\right)}\alpha\,\mathrm{sgn}\,(e)\,,\ \alpha_1, \alpha > 0 \tag{8.32}$$

$$\dot{b}_r = -\frac{q\underline{\tilde{w}}_r + (1-q)\bar{\tilde{w}}_r}{\left(q\underline{\tilde{w}}_r + (1-q)\bar{\tilde{w}}_r\right)^{\mathrm{T}}\left(q\underline{\tilde{w}}_r + (1-q)\bar{\tilde{w}}_r\right)}\left(\alpha_1 e + \alpha\,\mathrm{sgn}\,(e)\right) \tag{8.33}$$

$$\dot{q} = -\frac{1}{F(\underline{\widetilde{W}} - \overline{\widetilde{W}})^{\mathrm{T}}}\alpha\mathrm{sgn} \tag{8.34}$$

we have:

$$\begin{aligned}
\dot{y}_N \leq & -2\alpha\mathrm{sgn}(e) - \alpha_1 e - \alpha\,\mathrm{sgn}(e)\sum_{r=1}^{R}\sum_{i=1}^{I} x_i^2 \\
& + \kappa_0 + \kappa_1\|\dot{c}\| + \kappa_2\|\dot{\underline{\sigma}}\| + \kappa_3\|\dot{\bar{\sigma}}\| \\
= & -2\alpha\mathrm{sgn}(e) - \alpha\mathrm{sgn}(e)\sum_{r=1}^{R}\sum_{i=1}^{I} x_i^2 \\
& + \kappa_0 + \kappa_1\|\nu_c\| + \kappa_2\|\nu_{\underline{\sigma}}\| + \kappa_3\|\nu_{\bar{\sigma}}\|
\end{aligned}$$

in which:

$$\begin{aligned}
(M_1M_3 + M_2M_6)B_f &\leq \kappa_0 \\
(M_1M_4 + M_2M_7)B_f &\leq \kappa_1 \\
M_1M_5B_f &\leq \kappa_2 \\
N_2M_8B_f &\leq \kappa_3
\end{aligned}$$

The time derivative of the Lyapunov function is achieved as:

$$\begin{aligned}
\dot{V} \leq\ & e(-2\alpha\mathrm{sgn}(e) - \alpha_1 e - \alpha R\mathrm{sgn}(e)\sum_{i=1}^{I} x_i^2 \\
& + \kappa_0 + \kappa_1\|\nu_c\| + \kappa_2\|\nu_{\underline{\sigma}}\| + \kappa_3\|\nu_{\bar{\sigma}}\| - \dot{y}) \\
& + \sum_i\sum_k\left((\beta-1)\nu^2_{\underline{\sigma}_{ik}} + \alpha P_{\underline{\sigma}_{ik}}\nu_{\underline{\sigma}_{ik}} - \alpha_2\nu^4_{\underline{\sigma}_{ik}}\right) \\
& + \sum_i\sum_k\left((\beta-1)\nu^2_{\bar{\sigma}_{ik}} + \alpha P_{\bar{\sigma}_{ik}}\nu_{\bar{\sigma}_{ik}} - \alpha_2\nu^4_{\bar{\sigma}_{ik}}\right) \\
& + \sum_i\sum_k\left((\beta-1)\nu^2_{c_{ik}} + \alpha P_{c_{ik}}\nu_{c_{ik}} - \alpha_2\nu^4_{c_{ik}}\right)
\end{aligned}$$

From linear algebra, we have the following lemma:

Lemma 8.1. *For any positive real value λ and real values for a and b we have:*

$$|a||b| \leq \lambda a^2 + \lambda^{-1} b^2 \tag{8.35}$$

Considering Lemma 8.1, we get:

$$\begin{aligned}
\dot{V} \leq & -2\alpha|e| - \alpha_1 e^2 - \alpha R|e| \sum_{i=1}^{I} x_i^2 + (\lambda_1^{-1} + \lambda_2^{-1} + \lambda_3^{-1})e^2 \\
& + \kappa_0|e| + \kappa_1\lambda_1\|\nu_c\|^2 + \kappa_2\lambda_2\|\nu_{\underline{\sigma}}\|^2 + \kappa_3\lambda_3\|\nu_{\overline{\sigma}}\|^2 + B_{\dot{y}}|e| \\
& + \sum_i \sum_k \left((\beta - 1)\nu_{\underline{\sigma}_{ik}}^2 + \alpha P_{\underline{\sigma}_{ik}} \nu_{\underline{\sigma}_{ik}} - \alpha_2 \nu_{\underline{\sigma}_{ik}}^4\right) \\
& + \sum_i \sum_k \left((\beta - 1)\nu_{\overline{\sigma}_{ik}}^2 + \alpha P_{\overline{\sigma}_{ik}} \nu_{\overline{\sigma}_{ik}} - \alpha_2 \nu_{\overline{\sigma}_{ik}}^4\right) \\
& + \sum_i \sum_k \left((\beta - 1)\nu_{c_{ik}}^2 + \alpha P_{c_{ik}} \nu_{c_{ik}} - \alpha_2 \nu_{c_{ik}}^4\right)
\end{aligned}$$

in which λ_1, λ_2, and λ_3 have positive values. Considering the design variables α_1, α, and β as:

$$\lambda_1^{-1} + \lambda_2^{-1} + \lambda_3^{-1} \leq \alpha_1 \tag{8.36}$$

$$\kappa_0 + B_{\dot{y}} \leq \alpha \tag{8.37}$$

$$\beta \leq 1 - \max(2\kappa_1\lambda_1, 2\kappa_2\lambda_2, 2\kappa_3\lambda_3) \tag{8.38}$$

we have:

$$\begin{aligned}
\dot{V} \leq & -\alpha|e| - \alpha R|e| \sum_{i=1}^{I} x_i^2 \\
& + \sum_i \sum_k \left(\frac{\beta - 1}{2}\nu_{\underline{\sigma}_{ik}}^2 + \alpha(P_{\underline{\sigma}_{ik}}^2 + \nu_{\underline{\sigma}_{ik}}^2) - \alpha_2 \nu_{\underline{\sigma}_{ik}}^4\right) \\
& + \sum_i \sum_k \left(\frac{\beta - 1}{2}\nu_{\overline{\sigma}_{ik}}^2 + \alpha(P_{\overline{\sigma}_{ik}}^2 + \nu_{\overline{\sigma}_{ik}}^2) - \alpha_2 \nu_{\overline{\sigma}_{ik}}^4\right) \\
& + \sum_i \sum_k \left(\frac{\beta - 1}{2}\nu_{c_{ik}}^2 + \alpha(P_{c_{ik}}^2 + \nu_{c_{ik}}^2) - \alpha_2 \nu_{c_{ik}}^4\right)
\end{aligned}$$

It is further possible to choose β as $\beta \leq -(4\alpha + 1)$ so that we achieve the following equation for the time derivative of the Lyapunov function:

$$\begin{aligned}\dot{V} \leq & -\alpha|e| - \alpha R|e|\sum_{i=1}^{I} x_i^2 \\ & + \sum_i \sum_k \left(\frac{\beta - 1}{4} \nu_{\underline{\sigma}_{ik}}^2 + \alpha P_{\underline{\sigma}_{ik}}^2 - \alpha_2 \nu_{\underline{\sigma}_{ik}}^4 \right) \\ & + \sum_i \sum_k \left(\frac{\beta - 1}{4} \nu_{\overline{\sigma}_{ik}}^2 + \alpha P_{\overline{\sigma}_{ik}}^2 - \alpha_2 \nu_{\overline{\sigma}_{ik}}^4 \right) \\ & + \sum_i \sum_k \left(\frac{\beta - 1}{4} \nu_{c_{ik}}^2 + \alpha P_{c_{ik}}^2 - \alpha_2 \nu_{c_{ik}}^4 \right)\end{aligned}$$

The above equation implies the time derivative of the Lypapunov function is negative until the parameters of the T2FNN converge to a neighborhood near zero in which:

$$\begin{aligned}|\nu_{c_{ik}}| &\leq \frac{4\alpha}{1-\beta} P_{c_{ik}} \\ |\nu_{\underline{\sigma}_{ik}}| &\leq \frac{4\alpha}{1-\beta} P_{\underline{\sigma}_{ik}} \\ |\nu_{\overline{\sigma}_{ik}}| &\leq \frac{4\alpha}{1-\beta} P_{\overline{\sigma}_{ik}}\end{aligned}$$

and hence all the parameters of T2FNN are bounded.

8.4.2 Proposed Hybrid Training Algorithm for T2FNN With Type-2 MFs with Uncertain *c*

8.4.2.1 *T2FNN with Type-2 MFs with Uncertain c*

The upper and lower type-2 fuzzy Gaussian MFs with an uncertain center (Fig. 8.4) are presented as follows:

$$\overline{\mu}_{ik}(x_i) = \begin{cases} \exp\left(-\frac{1}{2} \frac{(x_i - \underline{c}_{ik})^2}{\sigma_{ik}^2} \right) & x_i < \underline{c}_{ik} \\ 1 & \underline{x}_{ik} \leq x_i \leq \overline{c}_{ik} \\ \exp\left(-\frac{1}{2} \frac{(x_i - \overline{c}_{ik})^2}{\sigma_{ik}^2} \right) & \overline{c}_{ik} < x_i \end{cases} \tag{8.39}$$

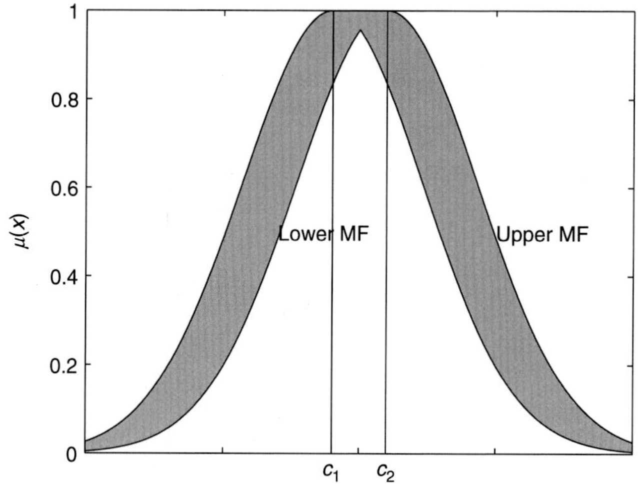

Figure 8.4 Type-2 Gaussian fuzzy MF with uncertain standard deviation.

$$\overline{\mu}_{ik}(x_i) = \begin{cases} \exp\left(-\frac{1}{2}\frac{(x_i - \overline{c}_{ik})^2}{\sigma_{ik}^2}\right) & x_i < \frac{\underline{c}_{ik}+\overline{c}_{ik}}{2} \\ \exp\left(-\frac{1}{2}\frac{(x_i - \underline{c}_{ik})^2}{\sigma_{ik}^2}\right) & \frac{\underline{c}_{ik}+\overline{c}_{ik}}{2} \leq x_i \end{cases} \tag{8.40}$$

where c_{ik} has an interval value $[\underline{c}_{ik}, \overline{c}_{ik}]$ and is the center parameter of the kth type-2 fuzzy set for the ith input. The parameter σ_{ik} is the standard deviation for the MFs and has a certain value.

The firing strength of the rules, which is itself an interval value, is calculated using the *prod* t-norm operator as follows:

$$\begin{aligned} \underline{w}_r &= \underline{\mu}_{\tilde{A}1}(x_1) * \underline{\mu}_{\tilde{A}2}(x_2) * \cdots \underline{\mu}_{\tilde{A}I}(x_I) \\ \overline{w}_r &= \overline{\mu}_{\tilde{A}1}(x_1) * \overline{\mu}_{\tilde{A}2}(x_2) * \cdots \overline{\mu}_{\tilde{A}I}(x_I) \end{aligned} \tag{8.41}$$

The consequent part corresponding to each fuzzy rule is a linear combination of the inputs $x_1, x_2 \ldots x_I$. This linear function is called f_r and is defined as $f_r = b_r + \sum_{i=1}^{I} a_{ri}x_i$. The output of the network is calculated as follows:

$$y_N = q\sum_{r=1}^{R} f_r\tilde{\underline{w}}_r + (1-q)\sum_{r=1}^{R} f_r\tilde{\overline{w}}_r \tag{8.42}$$

where $\underline{\tilde{w}_r}$ and $\overline{\tilde{w}_r}$ are the normalized values of the lower and upper output signals from the second hidden layer of the network as follows:

$$\underline{\tilde{w}_r} = \frac{\underline{w_r}}{\sum_{i=1}^{R} \underline{w_r}} \text{ and } \overline{\tilde{w}_r} = \frac{\overline{w_r}}{\sum_{i=1}^{R} \overline{w_r}} \tag{8.43}$$

The following vectors can be specified:
$\underline{\widetilde{W}}(t) = \left[\underline{\widetilde{w_1}}(t)\ \underline{\widetilde{w_2}}(t) \ldots \underline{\widetilde{w_R}}(t)\right]^{\mathrm{T}}$, $\overline{\widetilde{W}}(t) = \left[\overline{\widetilde{w_1}}(t)\ \overline{\widetilde{w_2}}(t) \ldots \overline{\widetilde{w_R}}(t)\right]^{\mathrm{T}}$, and $F = [f_1\, f_2\ \ldots f_R]$

The following assumptions have been used in this investigation: The time derivative of both the input and output signals can be considered bounded:

$$|\dot{x}_i(t)| \leq B_{\dot{x}}, |x_i(t)| \leq B_x, \quad (i = 1 \ldots I) \quad \text{and} \quad |\dot{y}(t)| \leq B_{\dot{y}} \quad \forall t \tag{8.44}$$

where $B_{\dot{x}}$, B_x, and $B_{\dot{y}}$ are assumed to be some known positive constants. It is clear that the equations of T2FNN with uncertain center values are quite similar to those of T2FNN with an uncertain σ value. The main difference is that when the center parameter of a type-2 MF is an interval, the MFs become undifferentiable in three points.

8.4.2.2 Proposed Hybrid Training Algorithm for T2FNN with MFs with Uncertain Center Value

The proposed hybrid PSO and SMC training method is as follows:

$$\begin{aligned} \dot{\underline{c}}_{sl} &= \nu_{\underline{c}_{sl}} \\ \dot{\nu}_{\underline{c}_{sl}} &= (\beta - 1)\nu_{\underline{c}_{sl}} - \alpha \underline{c}_{sl} + \gamma_1 Pl_{\underline{c}_{sl}} + \gamma_2 Pg_{\underline{c}_{sl}} - \alpha_2 \nu^3_{\underline{c}_{sl}}, \quad l = 1, \ldots, n, \\ &s = 1, \ldots, N \end{aligned} \tag{8.45}$$

$$\begin{aligned} \dot{\bar{c}}_{sl} &= \nu_{\bar{c}_{sl}} \\ \dot{\nu}_{\bar{c}_{sl}} &= (\beta - 1)\nu_{\bar{c}_{sl}} - \alpha \bar{c}_{sl} + \gamma_1 Pl_{\bar{c}_{sl}} + \gamma_2 Pg_{\bar{c}_{sl}} - \alpha_2 \nu^3_{\bar{c}_{sl}}, \quad l = 1, \ldots, n, \\ &s = 1, \ldots, N \end{aligned} \tag{8.46}$$

$$\begin{aligned} \dot{\sigma}_{sl} &= \nu_{\sigma_{sl}} \\ \dot{\nu}_{\sigma_{sl}} &= (\beta - 1)\nu_{\sigma_{sl}} - \alpha \sigma_{sl} + \gamma_1 Pl_{\sigma_{sl}} + \gamma_2 Pg_{\sigma_l} - \alpha_2 \nu^3_{\sigma_{sl}}, \quad l = 1, \ldots, n, \\ &s = 1, \ldots, N \end{aligned} \tag{8.47}$$

$$\dot{a}_{ri_s} = -x_i \frac{q\underline{\tilde{w}}_r + (1-q)\bar{\tilde{w}}_r}{\left(q\underline{\tilde{w}}_r + (1-q)\bar{\tilde{w}}_r\right)^{\mathrm{T}}\left(q\underline{\tilde{w}}_r + (1-q)\bar{\tilde{w}}_r\right)} \alpha \operatorname{sgn}(e),$$

$$s = 1, \ldots, N, r = 1, \ldots, R \tag{8.48}$$

$$\dot{b}_{r_s} = -\frac{q_s\underline{\tilde{w}}_{r_s} + (1-q_s)\bar{\tilde{w}}_{r_s}}{\left(q_s\underline{\tilde{w}}_{r_s} + (1-q_s)\bar{\tilde{w}}_{r_s}\right)^{\mathrm{T}}\left(q_s\underline{\tilde{w}}_{r_s} + (1-q_s)\bar{\tilde{w}}_{r_s}\right)} \left(\alpha_1 e_s + \alpha \operatorname{sgn}(e_s)\right)$$

$$s = 1, \ldots, N, r = 1, \ldots, R \tag{8.49}$$

$$\dot{q}_s = -\frac{1}{F_s(\underline{\widetilde{W}}_s - \overline{\widetilde{W}}_s)^{\mathrm{T}}} \alpha \operatorname{sgn}(e_s), \quad s = 1, \ldots, N \tag{8.50}$$

In (8.45)-(8.50), e_s is the identification error corresponding to sth particle, n is the number of type-2 MFs with uncertain centers, subscript s refers to the sth particle, and subscript l refers to lth dimension. For example, σ_{sl} means the lth dimension of sth particle corresponding to σ value of MFs. The cost function $f(e_s(t))$ is defined as:

$$f(e_s(t)) = \frac{1}{2}\sum e_s^2(t), \quad s = 1, \ldots, N \tag{8.51}$$

where the limits of the summation depend on the window size decided by the programmer, and the best local experiences of each particle $Pl_{\bar{c}_s}$, $Pl_{\underline{c}_s}$, and Pl_{σ_s} are updated using this cost function:

$$\begin{aligned} Pl_{\bar{c}_s}, Pl_{\underline{c}_s}, \text{ and } Pl_{\sigma_s} &= \arg\ \min\{f(e(t)|Pl_{\bar{c}_s}(t^-), Pl_{\underline{c}_s}(t^-), Pl_{\sigma_s}(t^-)), f(e(t)|\bar{c}_s(t), \\ &\quad \underline{c}_s(t), \sigma_s(t))\} \\ Pg_{\bar{c}}, Pg_{\underline{c}}, \text{ and } Pg_{\sigma} &= \arg\ \min\{f(e(t)|Pg_{\bar{c}}(t^-), Pg_{\underline{c}}(t^-), Pg_{\sigma}(t^-)), f(e(t)|\bar{c}_s(t), \\ &\quad \underline{c}_s(t), \sigma_s(t))\} \quad \forall s \in N \end{aligned}$$

Theorem 8.2. *The adaptation laws of* (8.12)-(8.17) *guarantee that the parameters of T2FNN in the form of* (8.42) *remain bounded provided that:*

$$|\dot{x}_i(t)| \le B_{\dot{x}}, |x_i(t)| \le B_x, \ (i = 1 \ldots I) \text{ and } |\dot{y}(t)| \le B_{\dot{y}} \quad \forall t \tag{8.52}$$

where $B_{\dot{x}}$, B_x, and $B_{\dot{y}}$ are assumed to be some known positive constants.

8.4.2.3 Stability Analysis

Since the stability of each particle is independent of the others, the stability of one particle is considered. Furthermore, for brevity the s in the indices does not appear. Since $\|x\| \leq B_x$ and $\|\dot{x}\| \leq B_{\dot{x}}$, the time derivative of (8.43) has the following properties:

$$\sum_{r=1}^{R} \dot{\tilde{\underline{w}}}_r \leq B_3 + B_4\|\dot{c}\| + B_5\|\dot{\sigma}\| \tag{8.53}$$

$$\sum_{r=1}^{R} \dot{\tilde{\underline{w}}}_r \leq B_6 + B_7\|\dot{c}\| + B_8\|\dot{\sigma}\| \tag{8.54}$$

in which B_3, B_4, B_5, B_6, B_7, and B_8 are positive real values. By using the following Lyapunov function, the stability of a single particle is analyzed.

$$V = \frac{1}{2}e^2 + \underbrace{\frac{\alpha_1}{2}\underline{c}^{\mathrm{T}}\underline{c} + \frac{1}{2}\nu_{\underline{c}}^{\mathrm{T}}\nu_{\underline{c}}}_{I_1} + \underbrace{\frac{\alpha_1}{2}\bar{c}^{\mathrm{T}}\bar{c} + \frac{1}{2}\nu_{\bar{c}}^{\mathrm{T}}\nu_{\bar{c}}}_{I_2} + \underbrace{\frac{\alpha_1}{2}\sigma^{\mathrm{T}}\sigma + \frac{1}{2}\nu_{\sigma}^{\mathrm{T}}\nu_{\sigma}}_{I_3} \tag{8.55}$$

To reduce the calculations required for the Lyapunov function and its time derivative the following assumption is made:

$$I_1 = \frac{\alpha_1}{2}\underline{c}^{\mathrm{T}}\underline{c} + \frac{1}{2}\nu_{\underline{c}}^{\mathrm{T}}\nu_{\underline{c}} \tag{8.56}$$

Therefore, we have the following equation for the time derivative of I_1:

$$\dot{I}_1 = \alpha_1 \sum_i \sum_k \underline{c}_{ik}\nu_{\underline{c}_{ik}} + \sum_i \sum_k \nu_{\underline{c}_{ik}}\left((\beta - 1)\nu_{\underline{c}_{ik}} - \alpha_1\underline{c}_{ik} + \alpha P_{\underline{c}_{ik}} - \alpha_2\nu_{\underline{c}_{ik}}^3\right) \tag{8.57}$$

and further:

$$\dot{I}_1 = \sum_i \sum_k \left((\beta - 1)\nu_{\underline{c}_{ik}}^2 + \alpha P_{\underline{c}_{ik}} - \alpha_2\nu_{\underline{c}_{ik}}^4\right) \tag{8.58}$$

Similar to the calculation of $\dot{I}_1$, the following equations for $\dot{I}_2$ and $\dot{I}_3$ are obtained:

$$\dot{I}_2 = \sum_i \sum_k \left((\beta - 1)\nu_{\bar{c}_{ik}}^2 + \alpha P_{\bar{c}_{ik}} - \alpha_2\nu_{\bar{c}_{ik}}^4\right) \tag{8.59}$$

$$\dot{I}_3 = \sum_i \sum_k \left((\beta - 1)\nu_{\sigma_{ik}}^2 + \alpha P_{\sigma_{ik}} - \alpha_2 \nu_{\sigma_{ik}}^4\right) \tag{8.60}$$

The time derivative of the Lyapunov function is obtained as follows:

$$\dot{V} = \dot{e}e + \dot{I}_1 + \dot{I}_2 + \dot{I}_3 = e(\dot{y}_N - \dot{y}) + \dot{I}_1 + \dot{I}_2 + \dot{I}_3 \tag{8.61}$$

Taking the time derivative of (8.42), the following term can be obtained:

$$\begin{aligned} \dot{y}_N &= \dot{q}\sum_{r=1}^{R} f_r\underline{\tilde{w}}_r + q\sum_{r=1}^{R}(\dot{f}_r\underline{\tilde{w}}_r + f_r\dot{\underline{\tilde{w}}}_r) - \dot{q}\sum_{r=1}^{R} f_r\bar{\tilde{w}}_r \\ &+ (1-q)\sum_{r=1}^{R}(\dot{f}_r\bar{\tilde{w}}_r + f_r\dot{\bar{\tilde{w}}}_r) \end{aligned} \tag{8.62}$$

By using (8.22), (8.53), and (8.62), the following can be obtained:

$$\begin{aligned} \dot{y}_N &= \dot{q}\sum_{r=1}^{R} f_r\underline{\tilde{w}}_r - \dot{q}\sum_{r=1}^{R} f_r\bar{\tilde{w}}_r + q\sum_{r=1}^{R}\left(\dot{f}_r\underline{\tilde{w}}_r + f_r(-\underline{\tilde{w}}_r\underline{K}_r + \underline{\tilde{w}}_r\sum_{r=1}^{R}\underline{\tilde{w}}_r\underline{K}_r)\right) \\ &+ (1-q)\sum_{r=1}^{R}\left(\dot{f}_r\bar{\tilde{w}}_r + f_r(-\bar{\tilde{w}}_r\overline{K}_r + \bar{\tilde{w}}_r\sum_{r=1}^{R}\bar{\tilde{w}}_r\overline{K}_r)\right) \\ &\leq \dot{q}\sum_{r=1}^{R}\left(f_r(\underline{\tilde{w}}_r - \bar{\tilde{w}}_r)\right) + \sum_{r=1}^{R}\left(\dot{f}_r(q\underline{\tilde{w}}_r + (1-q)\bar{\tilde{w}}_r)\right) \\ &+ B_f(B_1 + B_4) + B_fB_2\|\dot{\underline{c}}\| + B_fB_5\|\dot{\bar{c}}\| + B_f(B_3 + B_6)\|\dot{\sigma}\| \end{aligned} \tag{8.63}$$

Considering the adaptation laws as:

$$\dot{a}_{ri} = -x_i\frac{q\underline{\tilde{w}}_r + (1-q)\bar{\tilde{w}}_r}{\left(q\underline{\tilde{w}}_r + (1-q)\bar{\tilde{w}}_r\right)^{\mathrm{T}}\left(q\underline{\tilde{w}}_r + (1-q)\bar{\tilde{w}}_r\right)}\alpha\,\mathrm{sgn}\,(e) \tag{8.64}$$

$$\dot{b}_r = -\frac{q\underline{\tilde{w}}_r + (1-q)\bar{\tilde{w}}_r}{\left(q\underline{\tilde{w}}_r + (1-q)\bar{\tilde{w}}_r\right)^{\mathrm{T}}\left(q\underline{\tilde{w}}_r + (1-q)\bar{\tilde{w}}_r\right)}\left(\alpha_1 e + \alpha\,\mathrm{sgn}\,(e)\right) \tag{8.65}$$

$$\dot{q} = -\frac{1}{F(\underline{\widetilde{W}} - \overline{\widetilde{W}})^{\mathrm{T}}}\alpha\,\mathrm{sgn} \tag{8.66}$$

we have:

$$\begin{aligned}\dot{y}_N \leq & -2\alpha \mathrm{sgn}(e) - \alpha_1 e - \alpha \mathrm{sgn}(e) \sum_{r=1}^{R} \sum_{i=1}^{I} x_i^2 \\ & + \kappa_0 + \kappa_1 \|\dot{\sigma}\| + \kappa_2 \|\dot{\underline{c}}\| + \kappa_3 \|\dot{\overline{c}}\| \\ = & -2\alpha \mathrm{sgn}(e) - \alpha R \mathrm{sgn}(e) \sum_{i=1}^{I} x_i^2 \\ & + \kappa_0 + \kappa_1 \|\nu_\sigma\| + \kappa_2 \|\nu_{\underline{c}}\| + \kappa_3 \|\nu_{\overline{c}}\|\end{aligned}$$

in which:

$$\begin{aligned}(B_1 + B_4) B_f & \leq \kappa_0 \\ (B_3 + B_6) B_f & \leq \kappa_1 \\ B_2 B_f & \leq \kappa_2 \\ B_5 B_f & \leq \kappa_3\end{aligned}$$

The time derivative of the Lyapunov function is achieved as:

$$\begin{aligned}\dot{V} \leq\ & e(-2\alpha \mathrm{sgn}(e) - \alpha_1 e - \alpha R \mathrm{sgn}(e) \sum_{i=1}^{I} x_i^2 \\ & + \kappa_0 + \kappa_1 \|\nu_\sigma\| + \kappa_2 \|\nu_{\underline{c}}\| + \kappa_3 \|\nu_{\overline{c}}\| - \dot{y}) \qquad (8.67) \\ & + \sum_i \sum_k \left((\beta - 1)\nu_{\underline{c}_{ik}}^2 + \alpha P_{\underline{c}_{ik}} \nu_{\underline{c}_{ik}} - \alpha_2 \nu_{\underline{c}_{ik}}^4\right) \\ & + \sum_i \sum_k \left((\beta - 1)\nu_{\overline{c}_{ik}}^2 + \alpha P_{\overline{c}_{ik}} \nu_{\overline{c}_{ik}} - \alpha_2 \nu_{\overline{c}_{ik}}^4\right) \\ & + \sum_i \sum_k \left((\beta - 1)\nu_{\sigma_{ik}}^2 + \alpha P_{\sigma_{ik}} \nu_{\sigma_{ik}} - \alpha_2 \nu_{\sigma_{ik}}^4\right) \qquad (8.68)\end{aligned}$$

Considering Lemma 8.1 as is in (8.35) we get:

$$\begin{aligned}\dot{V} \leq & -2\alpha|e| - \alpha_1 e^2 - \alpha R |e| \sum_{i=1}^{I} x_i^2 + (\lambda_1^{-1} + \lambda_2^{-1} + \lambda_3^{-1}) e^2 \\ & + \kappa_0 |e| + \kappa_1 \lambda_1 \|\nu_\sigma\|^2 + \kappa_2 \lambda_2 \|\nu_{\underline{c}}\|^2 + \kappa_3 \lambda_3 \|\nu_{\overline{c}}\|^2 + B_{\dot{y}} |e|\end{aligned}$$

$$
\begin{aligned}
&+\sum_i\sum_k\left((\beta-1)\nu^2_{\underline{c}_{ik}}+\alpha P_{\underline{c}_{ik}}\nu_{\underline{c}_{ik}}-\alpha_2\nu^4_{\underline{c}_{ik}}\right)\\
&+\sum_i\sum_k\left((\beta-1)\nu^2_{\bar{c}_{ik}}+\alpha P_{\bar{c}_{ik}}\nu_{\bar{c}_{ik}}-\alpha_2\nu^4_{\bar{c}_{ik}}\right)\\
&+\sum_i\sum_k\left((\beta-1)\nu^2_{\sigma_{ik}}+\alpha P_{\sigma_{ik}}\nu_{\sigma_{ik}}-\alpha_2\nu^4_{\sigma_{ik}}\right)
\end{aligned}
\tag{8.69}
$$

in which λ_1, λ_2, and λ_3 has positive values. Considering the design variables α_1, α, and β as:

$$
\begin{aligned}
\lambda_1^{-1}+\lambda_2^{-1}+\lambda_3^{-1}&\leq\alpha_1\\
\kappa_0+B_{\dot{y}}&\leq\alpha\\
\beta\leq 1-\max(2\kappa_1\lambda_1&,2\kappa_2\lambda_2,2\kappa_3\lambda_3)
\end{aligned}
$$

we have:

$$
\begin{aligned}
\dot{V}\leq&-\alpha|e|-\alpha R|e|\sum_{i=1}^{I}x_i^2\\
&+\sum_i\sum_k\left(\frac{\beta-1}{2}\nu^2_{\underline{c}_{ik}}+\alpha(P^2_{\underline{c}_{ik}}+\nu^2_{\underline{c}_{ik}})-\alpha_2\nu^4_{\underline{c}_{ik}}\right)\\
&+\sum_i\sum_k\left(\frac{\beta-1}{2}\nu^2_{\bar{c}_{ik}}+\alpha(P^2_{\bar{c}_{ik}}+\nu^2_{\bar{c}_{ik}})-\alpha_2\nu^4_{\bar{c}_{ik}}\right)\\
&+\sum_i\sum_k\left(\frac{\beta-1}{2}\nu^2_{\sigma_{ik}}+\alpha(P^2_{\sigma_{ik}}+\nu^2_{\sigma_{ik}})-\alpha_2\nu^4_{\sigma_{ik}}\right)
\end{aligned}
$$

It is further possible to choose β as $\beta\leq-(4\alpha+1)$, so that we achieve the following equation for the time derivative of the Lyapunov function:

$$
\begin{aligned}
\dot{V}\leq&-\alpha|e|-\alpha R|e|\sum_{i=1}^{I}x_i^2\\
&+\sum_i\sum_k\left(\frac{\beta-1}{4}\nu^2_{\underline{c}_{ik}}+\alpha P^2_{\underline{c}_{ik}}-\alpha_2\nu^4_{\underline{c}_{ik}}\right)
\end{aligned}
$$

$$+\sum_i \sum_k \left(\frac{\beta - 1}{4} \nu_{\bar{c}_{ik}}^2 + \alpha P_{\bar{c}_{ik}}^2 - \alpha_2 \nu_{\bar{c}_{ik}}^4 \right)$$

$$+\sum_i \sum_k \left(\frac{\beta - 1}{4} \nu_{\sigma_{ik}}^2 + \alpha P_{\sigma_{ik}}^2 - \alpha_2 \nu_{\sigma_{ik}}^4 \right)$$

The above equation implies that the time derivative of the Lypapunov function is negative until the parameters of the T2FNN converge to a neighborhood near zero in which:

$$|\nu_{\sigma_{ik}}| \leq \frac{4\alpha}{1-\beta} P_{\sigma_{ik}}$$

$$|\nu_{\underline{c}_{ik}}| \leq \frac{4\alpha}{1-\beta} P_{\underline{c}_{ik}}$$

$$|\nu_{\bar{c}_{ik}}| \leq \frac{4\alpha}{1-\beta} P_{\bar{c}_{ik}}$$

and hence all of the parameters of T2FNN are bounded.

8.4.3 Further Discussion About the Hybrid Training Method

Since swarm-based optimization algorithms benefit from multiple solutions to the problem and simple update rules for the particles, they are used in combination with the SMC-based training algorithm to estimate the parameters of T2FNNs. The CPSO is used to estimate the parameters of the antecedent part and the SMC-based training method is used to estimate the parameters of the consequent part. The hybrid use of PSO for the training of the parameters of the antecedent part and SMC-based training algorithm for the parameters of the consequent part parameters has the following advantages over other training methods for T2FNNs:

- CPSO benefits from multiple solutions of the optimization problem, and it has less possibility of entrapment in local minima.
- Since CPSO benefits from simple rules, it may be a preferable choice with respect to GD and its variants, which necessitate the calculation of derivatives of output of FNN with respect to its parameters, because these parameters appear nonlinearly and their calculation is a difficult task.
- Another benefit of CPSO with respect to GD and its variants is that the selection of learning rate in these algorithms is very important because

a large value for the learning rate may cause instability while a small learning rate highly increases the possibility of the entrapment of the training algorithm in local minima.

- Unlike most of training methods which suffer from the lack of rigorous stability analysis, the stability analysis of the hybrid combination of CPSO and SMC-based training algorithm is fully considered using an appropriate Lyapunov function in this chapter.
- The hybrid combination of the CPSO and SMC-based training algorithm for the estimation of the parameters of T2FNN does not necessitate any matrix manipulations as is common with some of the training algorithms of T2FNN (e.g., extended KF and LM).

Furthermore, the proposed approach benefits from a new version of CPSO, called HCPSO, that has a nonlinear term and more degrees of freedom, which makes it possible to come up with better results. What is more, CPSO and HCPSO separately and/or in combination with the SMC-based training method are never used to train a FNN previously.

8.5 CONCLUSION

This chapter deals with a hybrid CPSO and SMC theory-based training method for the estimation of T2FNN parameters. Since CPSO benefits from a simple mathematical formulation, its use for the antecedent part parameters of T2FNNs makes it easier to train them. Moreover, since CPSO benefits from number of different solutions for the the antecedent part parameters, it is less probable that it will get stuck in local minima. The hybrid use of CPSO with SMC allows us have rigorous stability analysis using an appropriate Lyapunov function for the identification process.

REFERENCES

[1] C.-J. Lin, An efficient immune-based symbiotic particle swarm optimization learning algorithm for TSK-type neuro-fuzzy networks design, Fuzzy Sets Syst. 159 (21) (2008) 2890-2909.

[2] B. Allaoua, A. Laoufi, B. Gasbaoui, A. Abderrahmani, Neuro-fuzzy dc motor speed control using particle swarm optimization, Leonardo Electron. J. Pract. Tech. 15 (2009) 1-18.

[3] C.-J. Lin, S.-J. Hong, The design of neuro-fuzzy networks using particle swarm optimization and recursive singular value decomposition, Neurocomputing 71 (1) (2007) 297-310.

[4] H. Shayeghi, H. Shayanfar, PSO based neuro-fuzzy controller for LFC design including communication time delays, Int. J. Tech. Phys. Probl. Eng. 1 (2) (2010) 28-36.

[5] M.A. Shoorehdeli, M. Teshnehlab, A.K. Sedigh, M.A. Khanesar, Identification using ANFIS with intelligent hybrid stable learning algorithm approaches and stability analysis of training methods, Appl. Soft Comput. 9 (2) (2009) 833-850.
[6] M.A. Khanesar, M. Teshnehlab, E. Kayacan, O. Kaynak, A novel type-2 fuzzy membership function: Application to the prediction of noisy data, in: 2010 IEEE International Conference on Computational Intelligence for Measurement Systems and Applications (CIMSA), IEEE, 2010, pp. 128-133.
[7] M. Khanesar, M. Shoorehdeli, M. Teshnehlab, Hybrid training of recurrent fuzzy neural network model, in: International Conference on Mechatronics and Automation, 2007, ICMA 2007, IEEE, 2007, pp. 2598-2603.
[8] H.M. Emara, H.A.A. Fattah, Continuous swarm optimization technique with stability analysis, in: Proceedings of the 2004 American Control Conference, vol. 3, IEEE, 2004, pp. 2811-2817.

CHAPTER 9

Noise Reduction Property of Type-2 Fuzzy Neural Networks

Contents

Abstract

In this chapter, an attempt is made to show the effect of input noise in the rule base in a general way. There exist number of papers in literature claiming that the performance of T2FLSs is better than its type-1 counterparts under noisy conditions. We try to justify this claim by simulation studies only for some specific systems. However, in this chapter, such an analysis is done independent of the system to be controlled. For such an analysis, a novel type-2 fuzzy MF (elliptic MF) is proposed. This type-2 MF has certain values on both ends of the support and the kernel, and some uncertain values for other values of the support. The findings of the general analysis in this chapter and the aforementioned studies published in literature are coherent.

Keywords

Type-1 fuzzy neural networks, Type-2 fuzzy neural networks, TSK models, Artificial intelligence, Fuzzy logic, Neural networks

9.1 INTRODUCTION

In control theory, there is a direct relationship between the performance of the controller and the accuracy of the model of the system. What is more, in real life, it is always difficult to obtain a precise model to be controlled. Furthermore, the system is generally subjected to noise from both inside

http://dx.doi.org/10.1016/B978-0-12-802687-8.00009-8

and outside of the system. In such cases, the use of FLCs is preferable since they can handle lack of modeling and noise better than model-based controllers.

The performance of type-2 fuzzy sets in the presence of measurement noise has been considered in number of different papers. In many papers, it is concluded that T2FLSs give more promising control performance under noisy working environments [1]. For instance, in Ref. [2], the effect of measurement noise in type-1 and type-2 FLCs is simulated to perform comparative analysis of the responses of the systems in the presence of uncertainty. It is shown that the use of a T2FLC in real world applications which exhibit measurement noise can be preferable. In Ref. [3], type-2 fuzzy logic theory is applied to predict Mackey-Glass chaotic time-series with uniform noise. The comparison between type-1 and type-2 fuzzy systems shows the superiority of type-2 fuzzy systems in the presence of noise in the inputs.

A common claim in literature is that the performance of the T2FLSs is superior over their type-1 counterparts, especially under noisy conditions. However, the justifications offered for the claim made about the noise reduction property are generally limited to simulation studies carried out for specific systems. In this chapter, we propose a novel type-2 MF that enables us to come up with some metrics. The parameters of this function that represent uncertainty are de-coupled from the parameters that determine the center and support of the MF. This allows us to analyze the distortion of the rule base by the uncertainties in the inputs to the rule base. For this analysis, a simple T2FLS with the proposed novel MF is considered in which the effect of input noise in the rule base can be shown in a general way.

9.2 TYPE-2 FUZZY NEURAL SYSTEM STRUCTURE

9.2.1 Elliptic MF

A novel type-2 fuzzy MF is introduced. It has certain values on both ends of the support and kernel, and some uncertain values on other values of the support. The mathematical expression for the novel MF is expressed as:

$$\tilde{\mu}(x) = \begin{cases} \left(1 - \left|\frac{x-c}{d}\right|^{a}\right)^{\frac{1}{a}} & \text{if} \, c - d < x < c + d \\ 0 & \text{else} \end{cases} \tag{9.1}$$

where c and d are the center and the width of the MF and x is the input vector. The parameters a_1 and a_2 $(a_2 < a < a_1)$ determine the width of the uncertainty of the proposed MF, and these parameters should be selected as follows:

$$a_1 > 1 \qquad (9.2)$$
$$0 < a_2 < 1$$

Figures 9.1(a), (b), and (c) show the shapes of the proposed MF for $a_1 = a_2 = 1$, $a_1 = 1.2, a_2 = 0.8$, and $a_1 = 1.4, a_2 = 0.6$, respectively. As can be seen from Fig. 9.1(a), the shape of the proposed type-2 MF is changed to a type-1 triangular MF when its parameters are selected as $a_1 = a_2 = 1$. These parameters can be selected as some constants or they can be tuned adaptively.

9.2.2 Structure of the T2FLS

The interval T2FLS considered in this chapter benefits from type-2 MFs in the premise part and crisp numbers in the consequent part. Such a structure is called a A2-C0 fuzzy system in the literature [6], and the rule base is as follows:

$$\text{IF } x_1 \text{ is } \tilde{A}_{j1} \text{ and } x_2 \text{ is } \tilde{A}_{j2} \text{ and} \ldots \text{and } x_n \text{ is } \tilde{A}_{jn}$$
$$\text{THEN } u_j = \sum_{i=1}^{n} w_{ij} x_i + b_j \qquad (9.3)$$

where $x_1, x_2, \ldots, x_n$ are the input variables, $u_j (j = 1, \ldots, M)$ are the output variables, and $\tilde{A}_{ij}$ is a type-2 MFs for the jth rule and the ith input. w_{ij} and b_j $(i = 1, \ldots, n, j = 1, \ldots, M)$ are the parameters in the consequent part of the rules. The final output of the system can be written as [6]:

$$Y_{TSK/A2-C0} = \int_{f^1 \in [\underline{f}^1, \overline{f}^1]} \cdots \int_{f^M \in [\underline{f}^M, \overline{f}^M]} 1 \Big/ \frac{\sum_{j=1}^{M} f^j u_j}{\sum_{j=1}^{M} f^j} \qquad (9.4)$$

where $\underline{f}^j$ and $\overline{f}^j$ are given by:

$$\underline{f}^j(x) = \underline{\mu}_{\tilde{F}_1^j}(x_1) * \cdots * \underline{\mu}_{\tilde{F}_n^j}(x_n) \qquad (9.5)$$
$$\overline{f}^j(x) = \overline{\mu}_{\tilde{F}_1^j}(x_1) * \cdots * \overline{\mu}_{\tilde{F}_n^j}(x_n)$$

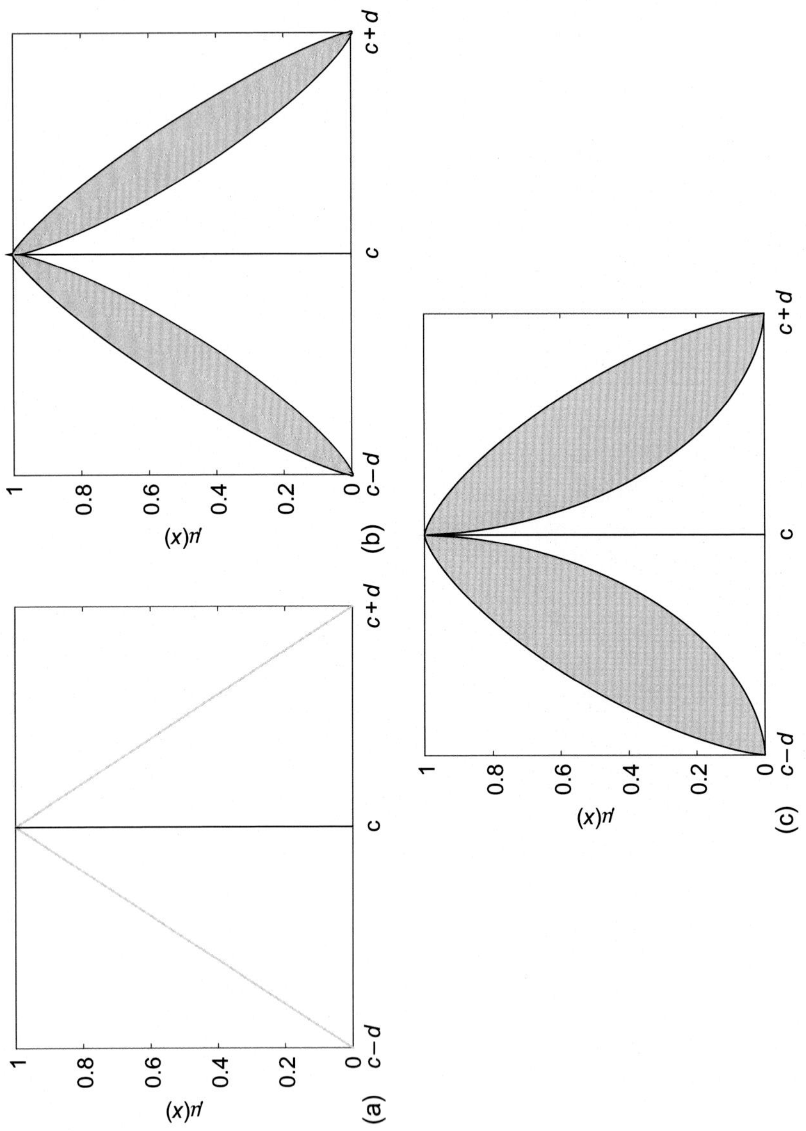

Figure 9.1 Shapes of the proposed type-2 MF with different values for a_1 and a_2.

in which $*$ represents the t-norm, which is the *prod* operator in this study. Although the exact computation of (9.4) requires Karnik-Mendel algorithm to be used, an approximate algorithm to compute the output of this fuzzy system in closed form is achieved by [6, 7]:

$$Y_{\text{TSK1}} = \frac{\sum_{j=1}^{M} \underline{f}^j u_j}{\sum_{j=1}^{M} \underline{f}^j + \sum_{j=1}^{M} \overline{f}^j} + \frac{\sum_{j=1}^{M} \overline{f}^j u_j}{\sum_{j=1}^{M} \underline{f}^j + \sum_{j=1}^{M} \overline{f}^j} \tag{9.6}$$

$$Y_{\text{TSK1}} = \frac{\sum_{j=1}^{M} (\underline{f}^j + \overline{f}^j) u_j}{\sum_{j=1}^{M} \underline{f}^j + \sum_{j=1}^{M} \overline{f}^j} \tag{9.7}$$

In this way, the firing of each rule is defined as follows:

$$r_j = \frac{\underline{f}^j + \overline{f}^j}{\sum_{j=1}^{M} \underline{f}^j + \sum_{j=1}^{M} \overline{f}^j} \tag{9.8}$$

9.2.3 Noise Reduction Property of the Proposed Type-2 MF

In this chapter, effort is made to prove the noise reduction property of T2FLSs. We consider two different fuzzy logic systems with the proposed novel MF. The first one is a T2FLS with one input with two MFs, and the other one is with two inputs with two MFs for each. As the parameters responsible for the width of uncertainty and the parameters responsible for the center and the support of the proposed MF (c and d, respectively) are decoupled from each other in the type-2 MF, it is possible to analyze how the width of uncertainty of the MF and the DCN in the rule base of the fuzzy system are related.

9.2.3.1 Case I

Let us consider a single input T2FLS that uses two MFs $\tilde{\mu}_1$ and $\tilde{\mu}_2$ such that:

$$\bar{\mu}_2 = 1 - \underline{\mu}_1 \tag{9.9}$$

$$\underline{\mu}_2 = 1 - \bar{\mu}_1 \tag{9.10}$$

The lower ($\underline{\mu}_1$) and upper ($\bar{\mu}_1$) MFs with the parameters c_1, d_1, a_1, and a_2 are defined as follows:

$$\bar{\mu}_1(x) = \begin{cases} \left(1 - |\frac{x-c_1}{d_1}|^{a_1}\right)^{1/a_1} & \text{if } c_1 - d_1 < x < c_1 + d_1 \\ 0 & \text{else} \end{cases} \tag{9.11}$$

$$\underline{\mu}_1(x) = \begin{cases} \left(1 - |\frac{x-c_1}{d_1}|^{a_2}\right)^{1/a_2} & \text{if } c_1 - d_1 < x < c_1 + d_1 \\ 0 & \textit{else} \end{cases} \tag{9.12}$$

The fuzzy system has two rules. Using (9.8), the firing strength of the first rule is calculated as:

$$r_1(x) = \frac{\bar{\mu}_1(x) + \underline{\mu}_1(x)}{2} \tag{9.13}$$

The firing strength will be distorted by the noise in the data as:

$$r_1(x+n) = \frac{\bar{\mu}_1(x+n) + \underline{\mu}_1(x+n)}{2} \tag{9.14}$$

where n indicates the noise added on the signal.

The total DCN over the support set can be found by the following integral:

$$\text{DCN} = \int_{n=-n_1}^{n=n_1} \int_{x=-d_1+c_1}^{x=d_1+c_1} [r_1(x) - r_1(x+n)]^2 \, \mathrm{d}x \, \mathrm{d}n \tag{9.15}$$

To simplify the limits in (9.15), the following definitions are done:

$$t = \frac{x - c_1}{d_1}, \; n' = \frac{n}{d_1} \tag{9.16}$$

Using the definitions in (9.16), the following equation is achieved:

$$\text{DCN} = d_1^2 \int_{n'=-n_1/d_1}^{n'=n_1/d_1} \int_{t=-1}^{t=1} [r_1(x) - r_1(x+n)]^2 \, \mathrm{d}t \, \mathrm{d}n' \tag{9.17}$$

In above, the parameter n_1 is the magnitude of the amplitude of the noise added onto the input of the fuzzy system.

The integral in (9.17) cannot be calculated explicitly. Therefore, a numerical solution of this integral is obtained for each pair of a_1 and a_2, and the distortion is drawn with respect to the a_1 and a_2 parameters. Figure 9.2 shows the numerical solution of the integral above for the noise level SNR $= 0$ dB. Note that to achieve SNR $= 0$ dB, n_1 is selected as being equal to d_1. As can be seen from the figure, the parameter a_2 is more critical in the noise reduction property of the proposed type-2 MF when compared to the parameter a_1. We can also see that the DCN in the case of $a_1 = a_2 = 1$, which corresponds to type-1 MF, is higher than the other values of a_1 and a_2, which correspond to the case of type-2 MFs. The figure indicates that by an appropriate selection of the parameters a_1 and a_2, it is possible to achieve better performance in the presence of noise. Although it has already been stated that the parameter a_1 should be selected bigger than 1 and the parameter a_2 between 0 and 1, Fig. 9.2 shows that there is a proper selection area for a_1 and a_2. Although the figure shows that it is better to select the parameter a_2 bigger than 0.3 and any value for a_1, a very small value for a_2 results in such a lower MF that the membership grade is

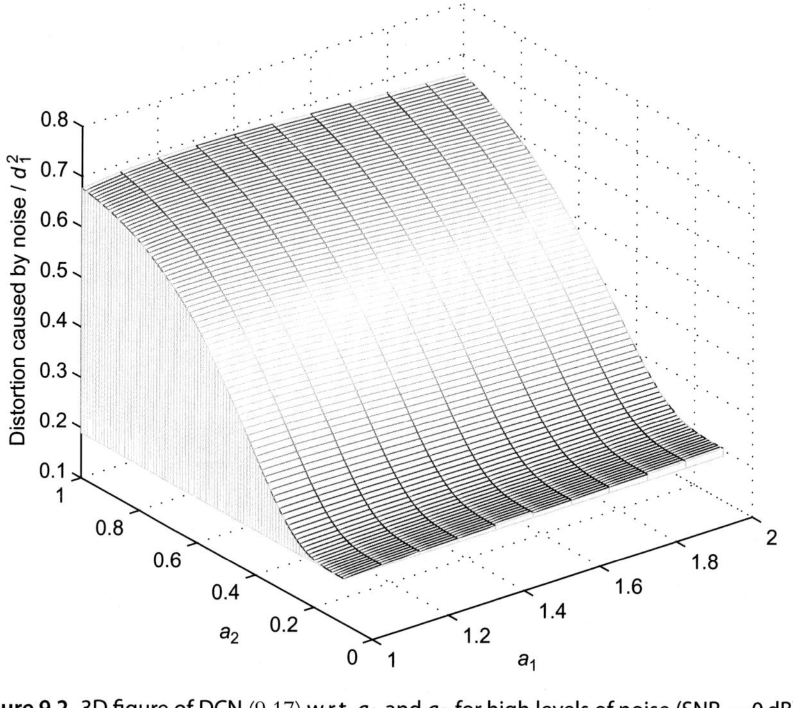

Figure 9.2 3D figure of DCN (9.17) w.r.t. a_1 and a_2 for high levels of noise (SNR $= 0$ dB).

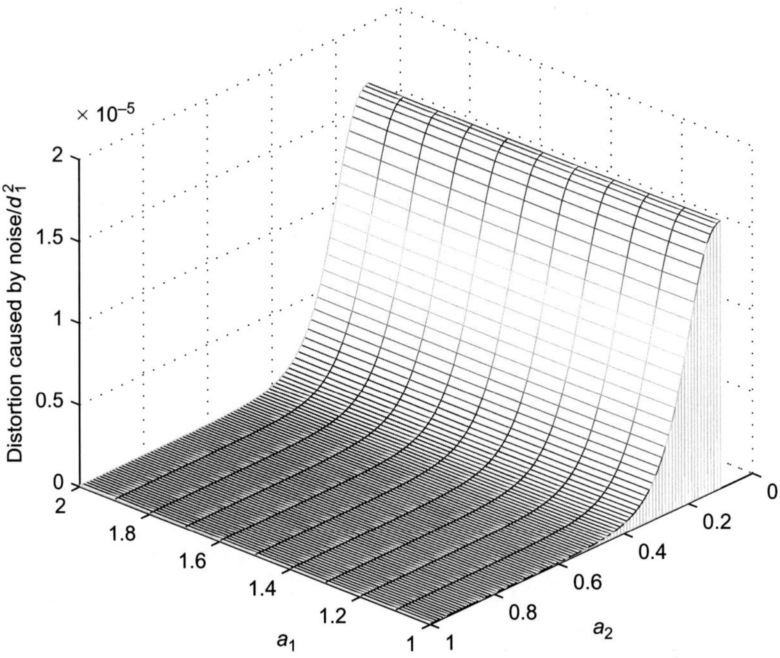

Figure 9.3 3D figure of DCN (9.17) w.r.t. a_1 and a_2 for low levels of noise (SNR = 100 dB).

close to zero for a large portion of its support. Similarly, a very large value for a_1 results in such an upper MF that the membership grade is close to one for a large portion of its support. Such MFs are not very desirable. An appropriate selection for a_1 and a_2 would therefore be $a_1 < 2$ and $a_2 > 0.5$.

Figure 9.3 shows the numerical solution of the integral for a low level of noise (SNR = 100 dB). This figure indicates that the noise reduction property of T1FLSs is comparable with T2FLSs in the presence of very low levels of noise.

9.2.3.2 Case II

Consider a T2FLS with two inputs and two type-2 MFs for each input. The MFs for the first input are selected as $\tilde{\mu}_{11}$ and $\tilde{\mu}_{12}$, and the MF selected for the second input are as $\tilde{\mu}_{21}$ and $\tilde{\mu}_{22}$. The type-2 fuzzy MF $\tilde{\mu}_{11}$ is defined as:

$$\bar{\mu}_{11}(x) = \begin{cases} \left(1 - |\frac{x-c_{11}}{d_{11}}|^{a_{1_{11}}}\right)^{1/a_{1_{11}}} & \text{if} |x - c_{11}| < d_{11} \\ 0 & \text{else} \end{cases} \tag{9.18}$$

$$\underline{\mu}_{11}(x) = \begin{cases} \left(1 - |\frac{x-c_{11}}{d_{11}}|^{a_{2_{11}}}\right)^{1/a_{2_{11}}} & \text{if} |x - c_{11}| < d_{11} \\ 0 & \text{else} \end{cases} \tag{9.19}$$

The type-2 fuzzy MF $\tilde{\mu}_{21}$ is defined as:

$$\bar{\mu}_{21}(x) = \begin{cases} \left(1 - |\frac{x-c_{21}}{d_{21}}|^{a_{1_{21}}}\right)^{1/a_{1_{21}}} & \text{if } |x - c_{21}| < d_{21} \\ 0 & \text{else} \end{cases} \tag{9.20}$$

$$\underline{\mu}_{21}(x) = \begin{cases} \left(1 - |\frac{x-c_{21}}{d_{21}}|^{a_{2_{21}}}\right)^{1/a_{2_{21}}} & \text{if } |x - c_{21}| < d_{21} \\ 0 & \text{else} \end{cases} \tag{9.21}$$

In order to see the effect of $a_{1_{11}}$ and $a_{2_{11}}$ on the DCN, $a_{1_{21}}$ and $a_{2_{21}}$ are set to 1. The other MFs are considered as:

$$\begin{aligned} \bar{\mu}_{12} &= 1 - \underline{\mu}_{11} \\ \underline{\mu}_{12} &= 1 - \bar{\mu}_{11} \\ \bar{\mu}_{22} &= 1 - \underline{\mu}_{21} \\ \underline{\mu}_{22} &= 1 - \bar{\mu}_{21} \end{aligned} \tag{9.22}$$

This fuzzy system has four rules. The firing strength for the first rule is written as:

$$r_1(x_1, x_2) = \frac{1}{2}\left(\bar{\mu}_{11}(x_1)\bar{\mu}_{21}(x_2) + \underline{\mu}_{11}(x_1)\underline{\mu}_{21}(x_2)\right) \tag{9.23}$$

The firing strength will be distorted by the noise in the data as:

$$r_1(x_1 + n, x_2) = \frac{1}{2}\left(\bar{\mu}_{11}(x_1 + n)\bar{\mu}_{21}(x_2) + \underline{\mu}_{11}(x_1 + n)\underline{\mu}_{21}(x_2)\right) \tag{9.24}$$

The total DCN over the support set can be calculated by the following integral:

$$\text{DCN} = \int_{n=-n_1}^{n=n_1} \int_{x_2=-d_{21}+c_{21}}^{x_2=d_{21}+c_{21}} \int_{x_1=-d_{11}+c_{11}}^{x_1=d_{11}+c_{11}} [r_1(x_1, x_2) - r_1(x_1 + n, x_2)]^2 \, \mathrm{d}x_1 \, \mathrm{d}x_2 \, \mathrm{d}n \tag{9.25}$$

To simplify the limits in (9.25), the following terms are defined:

$$t_1 = \frac{x_1 - c_{11}}{d_{11}},\ t_2 = \frac{x_2 - c_{21}}{d_{21}},\ n' = \frac{n}{d_{11}} \tag{9.26}$$

Using the definitions in (9.26), the following equation is achieved:

$$\mathrm{DCN} = d_{11}^2 d_{21} \int_{n'=-n_1/d_{11}}^{n'=n_1/d_{11}} \int_{t_2=-1}^{t_2=1} \int_{t_1=-1}^{t_1=1} [r_1(x_1, x_2) \\ -r_1(x_1 + n, x_2)]^2 \, \mathrm{d}t_1 \, \mathrm{d}t_2 \, \mathrm{d}n' \tag{9.27}$$

Similar to Case I, the integral in (9.27) cannot be calculated explicitly. Therefore, a numerical solution of this integral is obtained for each pair of a_1 and a_2, and the distortion is drawn with respect to the a_1 and a_2 parameters in Fig. 9.4. This figure is obtained with n_1 equal to d_1, which corresponds to SNR $=$ 0 dB. Figure 9.5, which is the contour diagram of Fig. 9.4, shows that there is an appropriate selection area for a_1 and a_2.

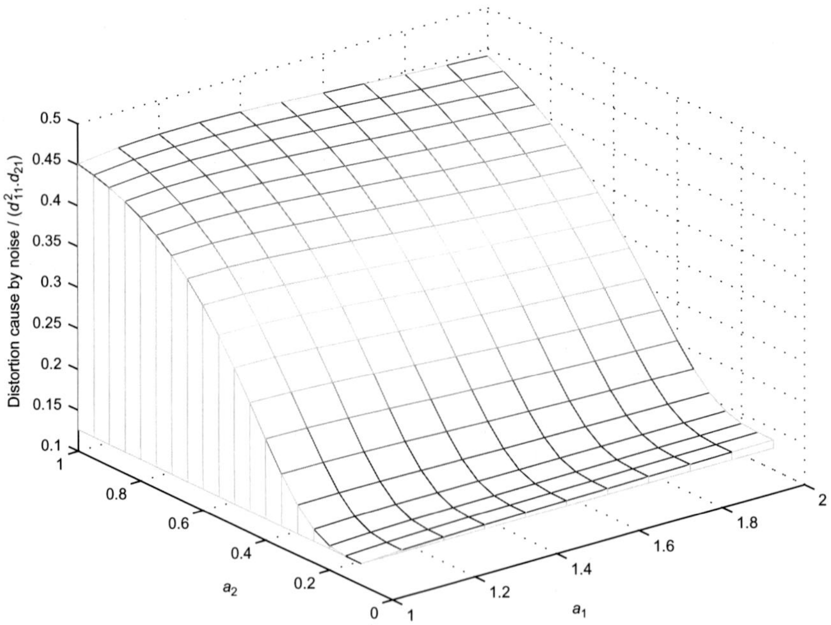

Figure 9.4 3D figure of DCN (9.27) w.r.t. a_1 and a_2 for high levels of noise (SNR $=$ 0 dB).

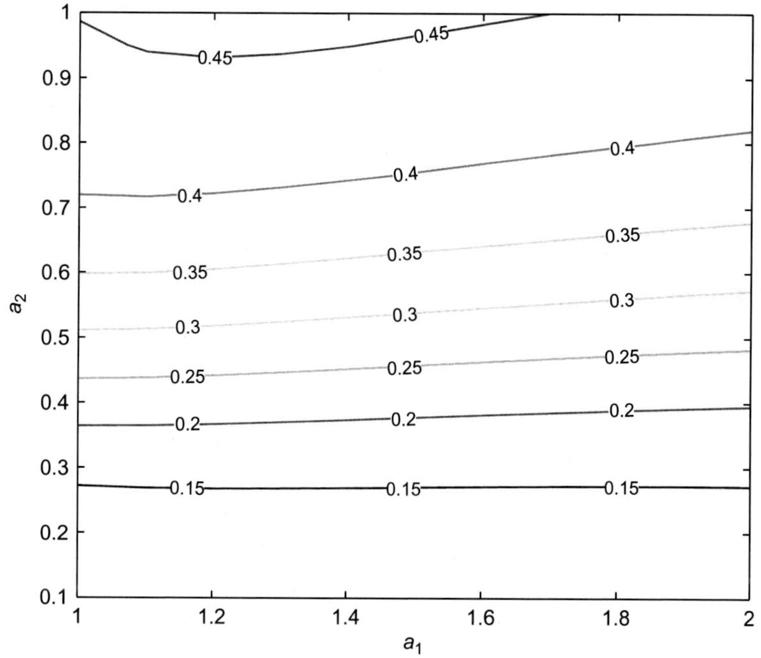

Figure 9.5 Contour figure of DCN (9.27) w.r.t. a_1 and a_2 (SNR = 0 dB).

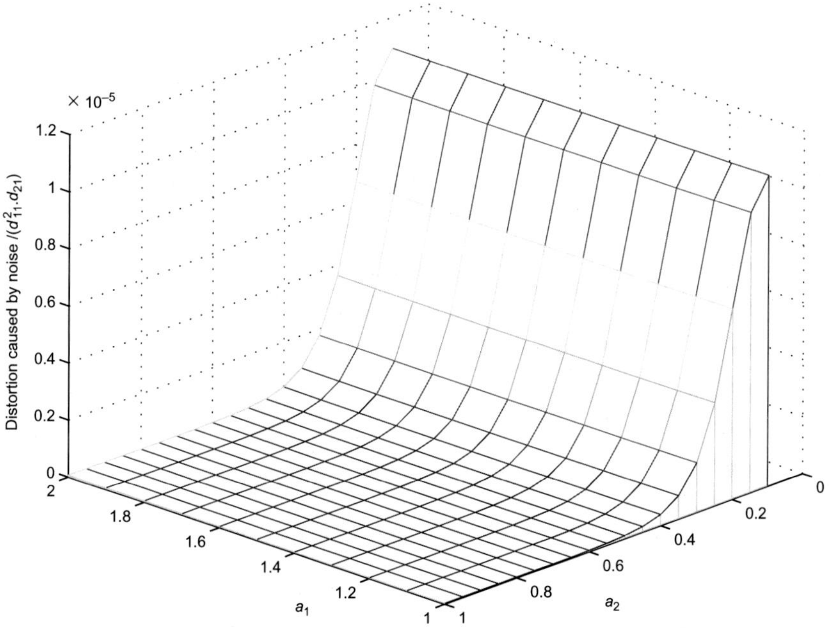

Figure 9.6 The 3D figure of distortion caused by noise (9.27) w.r.t. a_1 and a_2 for low levels of noise (SNR = 100 dB).

Based on similar arguments, an appropriate selection for a_2 and a_1 would therefore be $a_2 > 0.5$ and $a_1 < 2$.

Similar to Fig. 9.3, the numerical solution of the integral for a low level of noise (SNR = 100 dB) is shown in Fig. 9.6. This figure indicates that the noise reduction property of T1FLSs is comparable with T2FLSs in the presence of very low levels of noise. Therefore, in the presence of low levels of noise, T1FLS is preferable when compared to T2FLS because of its computational simplicity.

9.3 CONCLUSION

A novel type-2 fuzzy MF, elliptic type-2 MF, that has certain values on both ends of the support and the kernel and some uncertain values on other points of the support, is elaborated. With such a function, the parameters responsible for the width of uncertainty are de-coupled from the parameters responsible for the center and support of the MF. This allows us to analyze how the uncertainty in the input distorts the inference of the T2FLS.

REFERENCES

[1] C.-F. Juang, C.-H. Hsu, Reinforcement interval type-2 fuzzy controller design by online rule generation and Q-value-aided ant colony optimization, IEEE Trans. Syst. Man Cybernet. B Cybernet. 39 (2009) 1528-1542.

[2] R. Sepulveda, P. Melin, A. Rodriguez, A. Mancilla, O. Montiel, Analyzing the effects of the Footprint of Uncertainty in Type-2 Fuzzy Logic Controllers, Eng. Lett. 13 (2006) 138-147.

[3] J.M. Mendel, Uncertainty, fuzzy logic, and signal processing, Signal Process. 80 (2000) 913-933.

[4] M. Khanesar, E. Kayacan, M. Teshnehlab, O. Kaynak, Analysis of the noise reduction property of type-2 fuzzy logic systems using a novel type-2 membership function, IEEE Trans. Syst. Man Cybernet. B Cybernet. 41 (5) (2011) 1395-1406.

[5] J.M. Mendel, R.I.B. John, Type-2 fuzzy sets made simple, IEEE Trans. Fuzzy Syst. 10 (2002) 117-127.

[6] M. Begian, W. Melek, J. Mendel, Stability analysis of type-2 fuzzy systems, in: FUZZ-IEEE 2008 IEEE World Congress on Computational Intelligence, London, UK, 2008, pp. 947-953.

[7] J.M. Mendel, General Type-2 Fuzzy Logic Systems Made Simple: A Tutorial, in: Fuzzy Systems, IEEE Transactions on 22 (5) (2014) 1162-1182.

CHAPTER 10

Case Studies: Identification Examples

Contents

Abstract

In this chapter, the learning algorithms proposed in the previous chapters (GD-based, SMC theory-based, EKF and hybrid PSO-based learning algorithms) are used to identify and predict two nonlinear systems, namely Mackey-Glass and a second-order nonlinear time-varying plant. Several comparisons are made, and it has been shown that the proposed SMC theory-based algorithm has faster convergence than existing methods such as GD-based and swarm intelligence-based methods. Moreover, the proposed learning algorithm has an explicit form, and it is easier to implement than other existing methods. However, for offline algorithms for which computation time is not an issue, the hybrid training method based on PSO and SMC theory may be a preferable choice.

Keywords

Simulations, Mackey-Glass, Nonlinear time-varying system identification

10.1 IDENTIFICATION OF MACKEY-GLASS TIME SERIES

The first case study is the identification of a chaotic time series, namely the Mackey-Glass time series. This chaotic system is a well-known benchmark problem in literature described by the following dynamic equation [1]:

Fuzzy Neural Networks for Real Time Control Applications
http://dx.doi.org/10.1016/B978-0-12-802687-8.00010-4

$$\dot{x}(t) = 0.2\frac{x(t-\tau)}{1+x^{10}(t-\tau)} - 0.1x(t) \tag{10.1}$$

The numerical values selected for the chaotic system above are $\tau = 17$, $x(0) = 1.2$ in this case study. The first 118 data are eliminated. The predictor goal is to predict $x(t+6)$ using the inputs $x(t-18)$, $x(t-12)$, $x(t-6)$ and $x(t)$. For each input, two type-2 fuzzy Gaussian MFs with uncertain σ values are used. For the consequent part parameters, constant values are used. The number of rules in the system are therefore equal to 16. The number of training data is selected as 1000, and the number of test data is 200.

Figure 10.1(a) shows the RMSE versus the epoch number of SMC-theory based training algorithm, which indicates stable learning. As can be seen from this figure, the T2FNN gives accurate modeling results. In Fig. 10.1(b), the evolution of the adaptive learning rate is presented. Thanks to the stable adaptation law for this parameter, the learning rate converges to an appropriate value and there is no need for any trial and error stage for the selection of the learning rate. In Fig. 10.1(c), the correlation between the true output of the system and its estimated value is presented, which shows a satisfactory result. The histogram of the training error is depicted in Fig. 10.1(d). As can be seen from this figure, the probability density function of the error is almost Gaussian, which is highly desirable.

The results of applying different training algorithms for the prediction of Mackey-Glass are illustrated in Fig. 10.2(a). Moreover, the training error of different training algorithms as applied to the prediction of Mackey-Glass are presented in Fig. 10.2(b).

10.2 IDENTIFICATION OF SECOND-ORDER NONLINEAR TIME-VARYING PLANT

In the second case study, the proposed identification procedure is applied to a second-order nonlinear time-varying plant [2] described by the following equation:

$$y(k) = \frac{x_1(k)x_2(k) + x_3(k)}{x_4(k)} \tag{10.2}$$

where $x_1 = y(k-1)y(k-2)y(k-3)u(k-1)$, $x_2 = y(k-3) - b(k)$, $x_3 = c(k)u(k)$, and $x_4 = a(k) + y(k-2)^2 + y(k-3)^2$.

The time-varying parameters a, b and c in (10.2) are given by the following equations:

$$
\begin{aligned}
a(k) &= 1.2 - 0.2\cos(2\pi k/T) \\
b(k) &= 1 - 0.4\sin(2\pi k/T) \\
c(k) &= 1 + 0.4\sin(2\pi k/T)
\end{aligned}
\tag{10.3}
$$

where $T = 1000$ is the time span of the test.

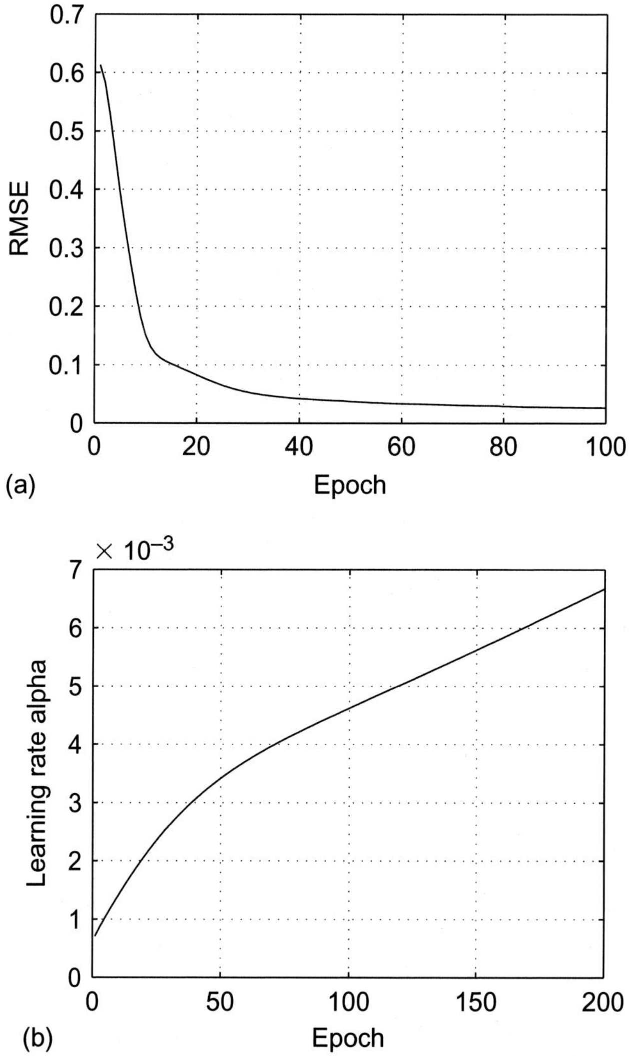

Figure 10.1 (a) RMSE of training versus epochs; (b) the evolution of the adaptive learning rate versus the epoch number; *(Continued)*

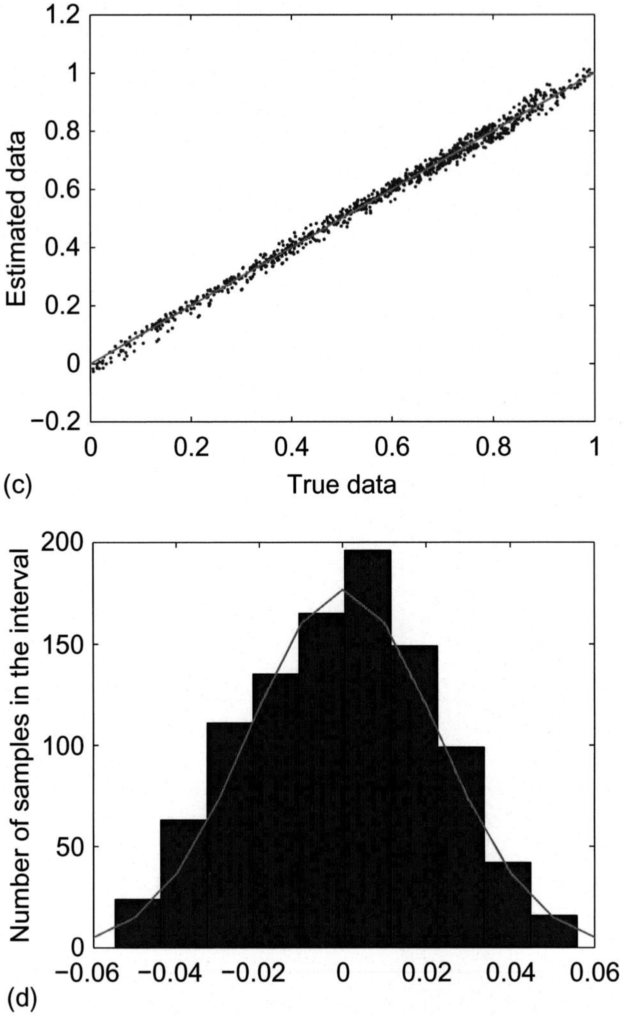

Figure 10.1, cont'd (c) the correlation analysis between the desired and the estimated values of T2FNN obtained by using SMC theory-based training algorithm; (d) the histogram of error obtained by using SMC theory-based learning algorithm.

Figure 10.3(a) shows the RMSE values versus epoch number, which indicates stable learning with the proposed learning algorithm. Thanks to the fully SMC theory-based parameter update rules as discussed in the previous chapters, the learning rate is also adaptive. The evolution of this parameter is illustrated in Fig. 10.3(b), and it is observed that the true value

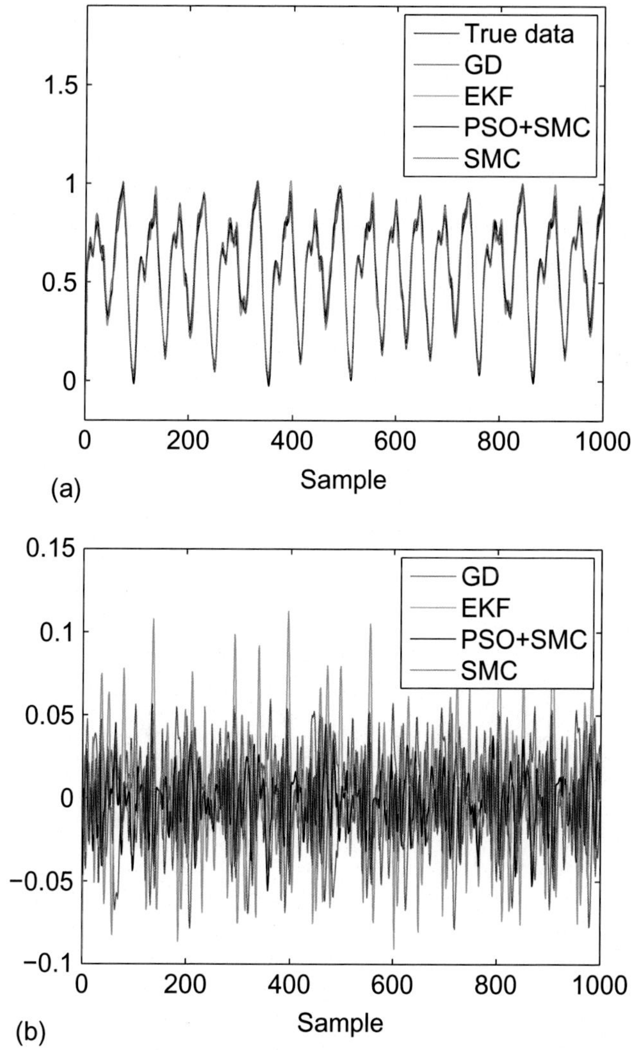

Figure 10.2 (a) The results obtained by using different training methods; (b) the error of training obtained by different training algorithms.

of the learning parameter is found automatically by the learning algorithm without any prior knowledge. Figure 10.3(c) shows the correlation between the true output of the system and its estimated value. Although there are some outliers in this figure, the overall performance is satisfactory. The histogram of the training error is depicted in Fig. 10.1(d). As can be seen

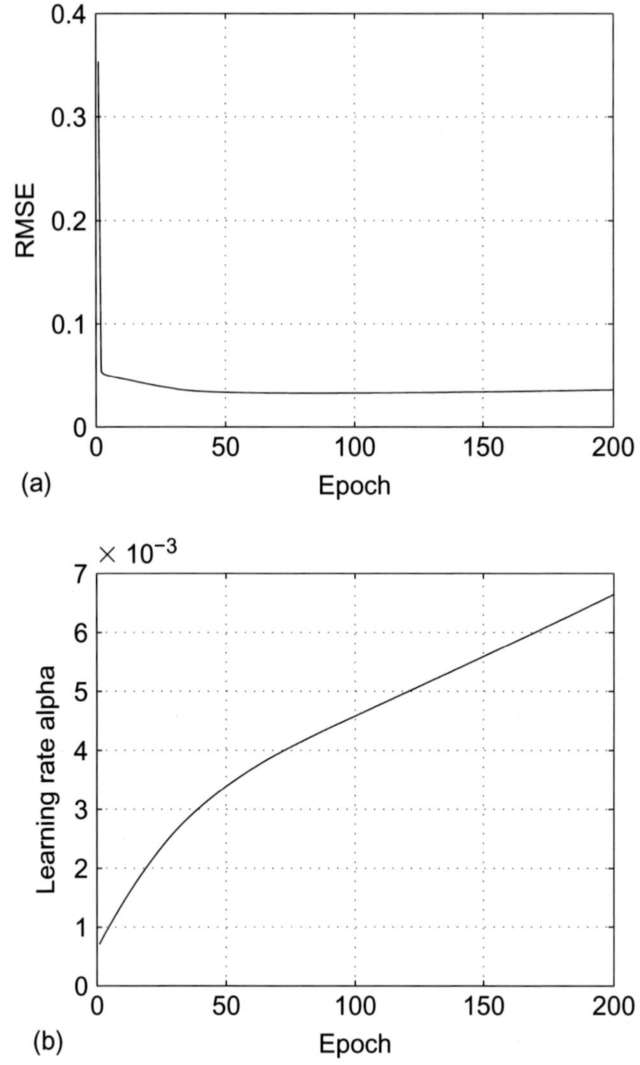

Figure 10.3 (a) RMSE of training versus epochs; (b) the evolution of the adaptive learning rate versus the epoch number;

from this figure, the probability density function of error is close to Gaussian, which is highly desirable.

The identification results and training error of different training algorithms to a time-varying nonlinear system are illustrated in Fig. 10.4(a) and (b), respectively.

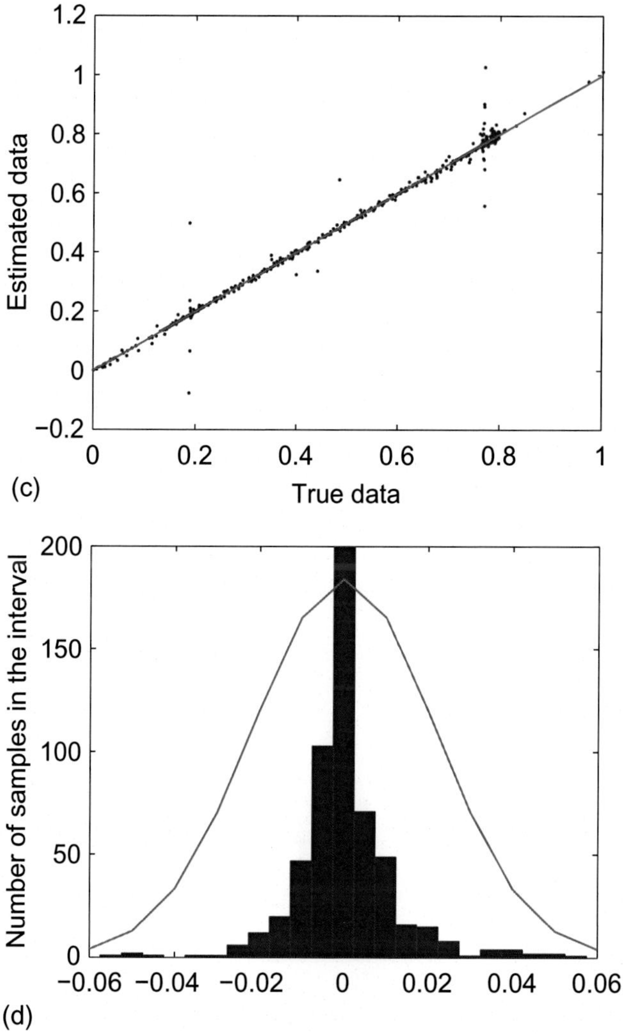

Figure 10.3, cont'd (c) the correlation analysis between the desired and the estimated values of T2FNN obtained by using SMC theory-based training algorithm; (d) the histogram of error obtained by using SMC theory-based learning algorithm.

10.3 ANALYSIS AND DISCUSSION

In order to have a numerical comparison between the RMSE values of different algorithms and their computation time, Tables 10.1 and 10.2

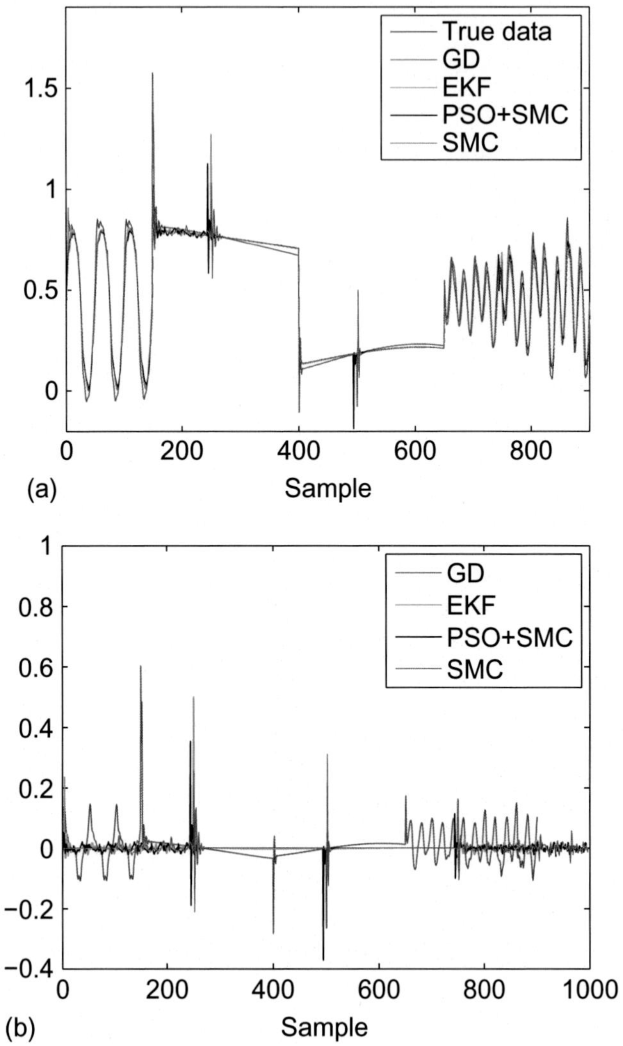

Figure 10.4 (a) Results obtained by using different training methods; (b) the error of training obtained by different training algorithms.

are presented. As can be seen from these two tables, the identification performances of the hybrid training method based on PSO and SMC theory are the best. However, since this algorithm needs different solutions to be run, it is the slowest. EKF and the proposed SMC theory-based learning algorithm are similar to each other, and they seem to be the best

Table 10.1 Comparison of different learning techniques

	Performance		
	Training	**Testing**	**Computation time (s)**
GD	0.0291	0.0250	55.79
EKF	0.0360	**0.0190**	61.4206
SMC	0.0224	0.0246	**52.65**
PSO+SMC	**0.0187**	0.0208	3039.25

Table 10.2 Comparison of different learning techniques

	Performance		
	Training	**Testing**	**Computation time (s)**
GD	0.0540	0.0613	124.1194
EKF	0.0275	**0.0261**	229.7074
SMC	0.0360	0.0390	**84.3878**
PSO+SMC	**0.0199**	0.0390	7086.78

when compared to other techniques. However, the computation time of the proposed SMC theory-based learning algorithm is significantly lower than the other methods. Moreover, the implementation of EKF requires the update of high-dimensional matrices and it is computationally very demanding. This conclusion results in the fact that although SMC theory-based learning algorithm may result in slight loss of performance, it is more practical in real-time applications.

As has been already mentioned, the reason for having the largest computation time for the EKF is that this algorithm includes the manipulation of high-dimensional matrices. A large amount of memory is used by these high-dimensional matrices, which makes the algorithm difficult to implement in most real-time applications. On the other hand, there are no matrix manipulations in the proposed SMC theory-based rules. Moreover, GD, LM, EKF and other gradient-based training methods include the calculations of the partial derivatives of the output with respect to the parameters, which is very difficult, and do not have any explicit form. The divergence issues are other problems of gradient-based training algorithms, which are solved by the SMC theory-based learning algorithm.

According to our observations, the overall most accurate algorithm regardless of the computation time, is the hybrid training method based on PSO and SMC theory. The basic idea behind this algorithm is that the parts that appear nonlinearly in the output can be trained by using HCPSO, while the parameters that appear linearly in the output can be trained using the SMC theory-based learning algorithm. This is because PSO benefits from a random optimization technique, and the use of it for the parameters of the consequent part reduces the speed of convergence considerably. However, the computational backbone of SMC theory-based learning algorithms makes them more logical than random optimization methods. This method is also implemented on a A2-C0 fuzzy model as was discussed in previous chapters. The results are summarized in Tables 10.1 and 10.2. As can be seen from these tables, as an offline training method, the hybrid training method based on PSO and SMC theory may be preferable. However, its computational time is higher than other training methods. The reason behind this computational burden is that the hybrid training method based on PSO and SMC theory necessitates too much feed-forward computation of T2FNN, which is a complex and time-consuming task.

10.4 CONCLUSION

The simulation results indicate the potential of the proposed SMC theory-based structure in real-time systems since the computation time of the proposed algorithm is significantly lower than that of the other methods, while keeping its high identification accuracy. Note that these parameter update rules can also be used for control purposes in which the computation time is prominent. One issue that should be taken into account is that the adaptation laws proposed in this chapter are continuous. However, for the simulation of the method in a computer, an optimal sampling time should be chosen. The choice of optimal sampling time may be a problem, because a very large value for the sampling time may cause instability in the system.

Moreover, for offline cases, where the computational time and memory consumption is not an issue, the hybrid training algorithm based on PSO and SMC theory is a preferable choice. The use of PSO as the optimizer for the parameters that appear nonlinearly in the output of T2FNN makes it possible to lessen the possibility of entrapment of the algorithm in a local minima and improves the performance of the results.

REFERENCES

[1] M.C. Mackey, L. Glass, et al., Oscillation and chaos in physiological control systems, Science 197 (4300) (1977) 287-289.
[2] R.H. Abiyev, O. Kaynak, Type 2 fuzzy neural structure for identification and control of time-varying plants, IEEE Trans. Indust. Electron., 57 (12) (2010) 4147-4159.

CHAPTER 11

Case Studies: Control Examples

Contents

Abstract

In this chapter, three real-world control problems, namely anesthesia, magnetic rigid spacecraft and tractor-implement system are studied by using SMC theory-based learning algorithms for T2FNNs. For all the systems, the FEL scheme is preferred in which a conventional controller (PD, etc.) works in parallel with an intelligent structure (T1FNN, T2FNN, etc.). The proposed learning algorithms have been shown to be able to control these real-world example problems with satisfactory performance. Note that the proposed control algorithms do not need a priori knowledge of the system to be controlled.

Keywords

Anesthesia, Elliptic membership function, Tractor and implement, Bispectral index

Fuzzy Neural Networks for Real Time Control Applications
http://dx.doi.org/10.1016/B978-0-12-802687-8.00011-6

11.1 CONTROL OF BISPECTRAL INDEX OF A PATIENT DURING ANESTHESIA

Over the last few decades, control engineering has widely influenced modern medicine. One of the most influenced fields of modern medicine is pharmacology. Closed-loop insulin delivery [1], closed-loop control of arterial pressure by infusion of sodium nitroprusside [2] and many more can be listed as the examples of the applications of control engineering in pharmacology. Closed-loop drug delivery for anesthesia [3, 4] is also one of the most important applications of control engineering in medicine. The first step to control the depth of anesthesia is to measure the depth of anesthesia. There are different methods to monitor the depth of anesthesia during surgery. One of the most well-known methods to monitor depth of anesthesia is called BIS. The use of different control methodologies in anesthesia makes it possible to use less volume of drug while maintaining the BIS of patient in an appropriate level for performing the surgery. Using less volume of drug means less side effects and less recovery time after surgery, which reduces the cost of the surgery.

The dynamical model of a patient is naturally uncertain since it depends on different parameters, e.g., gender, age, weight, height, and heart rate. For such an uncertain system, the use of model-free approaches, i.e., fuzzy systems and neural networks, is more preferable than model-based approaches. In this chapter, among the model-free approaches, T2FNNs are preferred to cope with high levels of uncertainties. Such a selection is more promising to control the BIS of a patient during anesthesia in which the model is highly uncertain and differs from one patient to another, and may even vary during the surgery for a single patient.

In this case study, the proposed novel FEL scheme is used to control BIS during anesthesia in which a T2FNN works in parallel with a PD controller (see Fig. 11.1). This control strategy has already been studied in Chapter 7. The T2FNN has two inputs: the error and the time derivative of the error. The MFs considered for the system are Gaussian type-2 MFs with uncertain variance. The SMC theory-based parameter update rules are derived for such a structure, and the stability of the learning algorithm was already considered in Chapter 7.

T2FLCs were previously applied to control BIS during anesthesia in Ref. [5]. However, in this study the T2FLC does not include any adaptation and hence this study is the first study in which BIS is controlled using T2FLS that results in an adaptive control scheme. Moreover, since the proposed

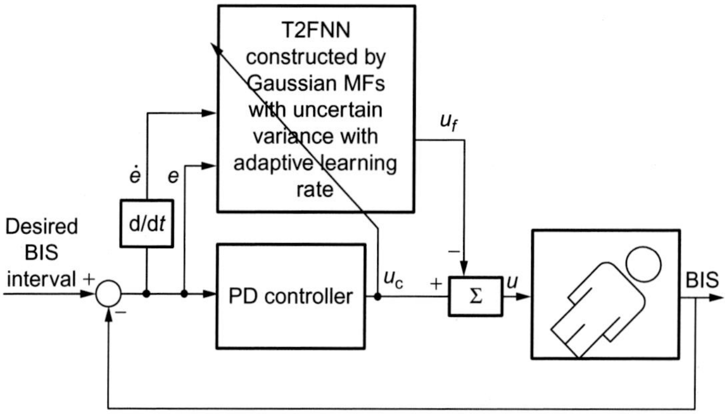

Figure 11.1 Block diagram of the proposed T2FNN scheme.

method uses type-2 MFs, it is expected that the system outperforms its type-1 counterpart especially when measurement noise is significant in the system. The simulation results show that the proposed approach can control BIS level during anesthesia in the presence of measurement noise and uncertainties that necessarily exist in the model of the patient. The performance of the T2FNN is also compared with that of its type-1 counterpart in the presence of noise. It is also shown that the states of the system follow the predefined sliding motion.

11.1.1 Realistic Patient Model

There exist several models that describe the dynamics of the patient when a specific drug (in this case Propofol) is injected. This relationship is described by pharmacokinetic and pharmacodynamic models. Pharmacokinetic models describe the distribution of drugs in the body and pharmacodynamic models represent the relationship between blood concentration of a drug and its clinical effects. The pharmacokinetics can be illustrated using a three-compartment model as follows [4, 6]:

$$
\begin{aligned}
\dot{x}_1(t) &= -(k_{10} + k_{12} + k_{13})x_1(t) + k_{21}x_2(t) + k_{31}x_3(t) \\
\dot{x}_2(t) &= k_{12}x_1(t) - k_{21}x_2(t) \\
\dot{x}_3(t) &= k_{13}x_1(t) - k_{31}x_3(t)
\end{aligned}
\tag{11.1}
$$

where $x_1(t)$ denotes the amount of drug in the central compartment. The peripheral compartments number two and number three model the drug exchange of the blood with well and poorly perfused body tissues. The amount of the blood in these compartments is represented by $x_2(t)$ and $x_3(t)$, respectively. The constants k_{ij} represent the transfer rate of the drug from the jth compartment to the ith compartment. The constant k_{10} is the rate of drug metabolism and $u(t)$ is the infusion rate of the Propofol into the blood. Other constants for the Schider model [6] for Propofol are as follows:

$$\begin{aligned}
&V_1 = 4.27[l],\ V_2 = 18.9 - 0.391(age - 53)[l],\\
&C_{l1} = 1.89 + 0.0456(weight - 77) - 0.0681(lbm - 59) + 0.0264(height - 177)\\
&C_{l2} = 1.29 - 0.024(age - 53),\ C_{l3} = 0.836,\ k_{10} = \frac{C_{l1}}{V_1},\ k_{12} = \frac{C_{l2}}{V_1},\\
&k_{13} = \frac{C_{l3}}{V_1},\ k_{21} = \frac{C_{12}}{V_2},\ k_{31} = \frac{C_{l3}}{V_3}
\end{aligned} \tag{11.2}$$

The lbm depends on gender: for a male person we have:

$$\text{lbm} = 1.1\,\text{weight} - 128\frac{\text{weight}^2}{\text{height}^2} \tag{11.3}$$

while for a female person we have:

$$\text{lbm} = 1.07\,\text{weight} - 148\frac{\text{weight}^2}{\text{height}^2} \tag{11.4}$$

The pharmacodynamics is characterized by a first-order differential equation as follows:

$$\dot{C}_e(t) = -0.456C_e(t) + 0.456C_p(t) \tag{11.5}$$

in which C_p is the concentration of the drug in the blood and is calculated as:

$$C_p = \frac{x_1}{V_1} \tag{11.6}$$

and C_e is the concentration of the drug in the effect compartment. In this chapter, the depth of anesthesia is monitored using BIS. The BIS can have any value from 0 to 100. If the BIS is equal to *zero*, the patient does not have any cerebral activity. On the other hand, if it is equal to 100, the patient is fully awake. BIS and C_e are related to each other by the following static equation:

$$\mathrm{BIS}(t) = E_0 - E_{\max}\frac{C_e^{\gamma}(t)}{C_e^{\gamma}(t) + C_{50}^{\gamma}} \tag{11.7}$$

in which E_0 is the BIS for the awake state and is typically taken equal to 100. $E_{\max}$ denotes the maximum effect of the drug, and C_{50} and γ are the drug concentration at half-maximal effect and the steepness of the curve, which vary from one patient to another. Also note that C_{50} depends on the heart rate and varies during surgery. The following equation illustrates the relation between C_{50} and HR [7]:

$$C_{50} = C_0 \log(HR) \tag{11.8}$$

in which C_0 is a constant value. This fact makes the patient model an uncertain nonlinear dynamic system.

11.1.2 Simulation Results

The parameters of the three compartment models considered in this chapter vary from one patient to another, or may be time-varying for a single patient. These values for 12 different patients are given in Table 11.1.

The state of a patient during surgery includes three phases: induction, maintenance and recovery. At the beginning of the induction phase, the injection of Propofol starts and the patient loses consciousness completely.

Table 11.1 Characteristic variables for each of the 12 patients used in this study

Patient	Age	Length	Weight	Gender	C_{50}	E_0	$E_{\max}$	γ
1	40	163	54	F	6.33	98.80	94.10	2.24
2	36	163	50	F	6.76	98.60	86.00	4.29
3	28	164	52	F	8.44	91.20	80.70	4.10
4	50	163	83	F	6.44	95.90	102.00	2.18
5	28	164	60	M	4.93	94.70	85.30	2.46
6	43	163	59	F	12.10	90.20	147.00	2.42
7	37	187	75	M	8.02	92.00	104.00	2.10
8	38	174	80	F	6.56	95.50	76.40	4.12
9	41	170	70	F	6.15	89.20	63.80	6.89
10	37	167	58	F	13.70	83.10	151.00	1.65
11	42	179	78	M	4.82	91.80	77.90	1.85
12	34	172	58	F	4.95	96.20	90.80	1.84

In the maintenance phase, in which the surgery is performed, the administration of Propofol continues and the patient is kept unconscious. Finally, at the end of the surgery, the injection of Propofol is stopped and some other drugs may be used to awaken the patient. In this chapter, only the first two stages are considered. The initial values of the parameters of the controller for different patients are considered to be the same. The MFs of the fuzzy controller are uniformly distributed over the support set and the parameters of the consequent part are taken to be equal to zero. As can be seen in Fig. 11.2, the desired operating region for a typical patient during a surgery with automated anesthesia is in the range of $[40,\ 60]$. This range is suggested by the manufacturers of BIS monitoring systems. It is further observed from Fig. 11.2 that any BIS level higher than 70 corresponds to light sedation and may cause the patient to unexpectedly wake up or experience pain. Moreover, any BIS level less than 30 is called deep anesthesia and should be prevented during surgery. BIS level is observed and controlled by an anesthetist during surgery. It has been shown that the mortality rate of the older people is 16.7% for an average BIS value less than 40 as compared to 4.2% for an average BIS less than 60. This study implies that keeping middle-aged and older patients in a deep hypnotic stage increases the possibility of mortality [8]. In this study, any BIS level less than 40 is considered to be

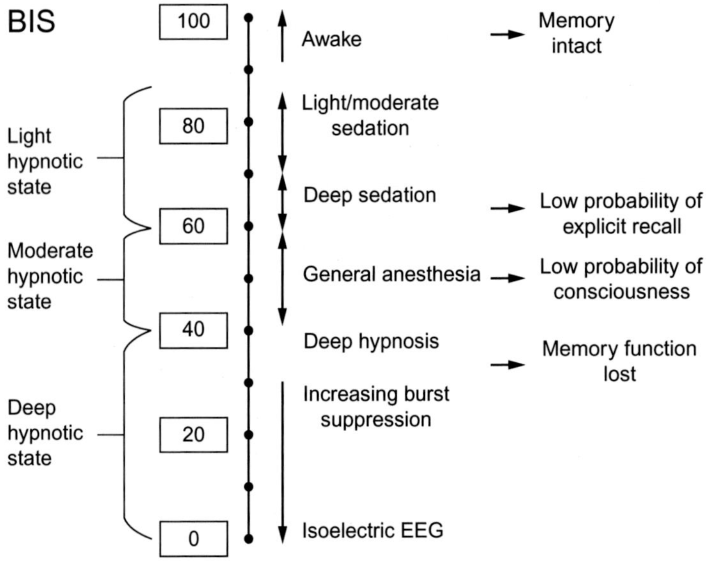

Figure 11.2 Meaning of different levels of BIS [9].

undesirable and is prevented during simulations. Furthermore, since it is impossible to eliminate the drug from the patient's body, the control signal cannot be negative. The injection rate of Propofol to the body of the patient also has an upper limit that is defined by the maximum injection rate of the automatic pump used during surgery and the maximum permissible injection rate that do not harm the patient. In this case study, the maximum value of the injection rate is considered to be 3.2 mg/s as in ref. [4].

The settling time during the induction phase is defined as the time it takes for the BIS of a patient to reach the interval of [45, 55] and to remain in this interval. The results obtained by applying an MPC to 12 patients is considered in a paper, which shows that the settling time varies from 90 *sec* to 190 *sec* and the BIS level of patient no. 10 falls down 45 during induction phase [4]. Moreover, the settling times for the condition when EPSC is used vary from 75 *sec* to 200 *sec* [4]. Furthermore, the results of applying EPSC show that the BIS of patients no. 5 and no. 12 fall down to 45 during induction, which is undesirable.

The results of the proposed controller for the 12 patients during the induction phase are shown in Fig. 11.3. As can be seen from Fig. 11.3, the settling times of the proposed algorithm vary from 240 *sec* to 742 *sec* and are longer than those of MPC and EPSC. However, the BIS does not fall to 45 for any patient, which is highly desirable. Another advantage of the proposed method over MPC and EPSC is that the control signals (the rates of injection of Propofol) do not meet their saturation value for any patients when applying MPC or EPSC (see Fig. 11.3). In summary, if the proposed method is used for the control of anesthesia, although it takes approximately 30 s to 11 min more time for a patient to become ready for surgery, the depth of the patient's anesthesia never becomes deeper than needed during the induction phase. In other words, by the use of the proposed method, the patient changes to the moderate hypnotic state slowly but more safely, and the maximum rate of drug injection to a patient never hits its upper bound.

The response of the proposed controller in the maintenance phase is shown in Figs. 11.4 and 11.5. In order to investigate the performance of the controller under more realistic conditions, it is assumed that the heart rate varies considerably every 5 min. As mentioned earlier, the variations in heart rate influence the parameter C_{50}. Thus, it is assumed that this parameter varies to a new value within 90% to 110% of its initial value every 5 min. From a control engineer point of view, the system would be an uncertain system considering these variations. As can be seen, the BIS level falls to 40 for a few seconds and is almost always within the desired range of [40, 60].

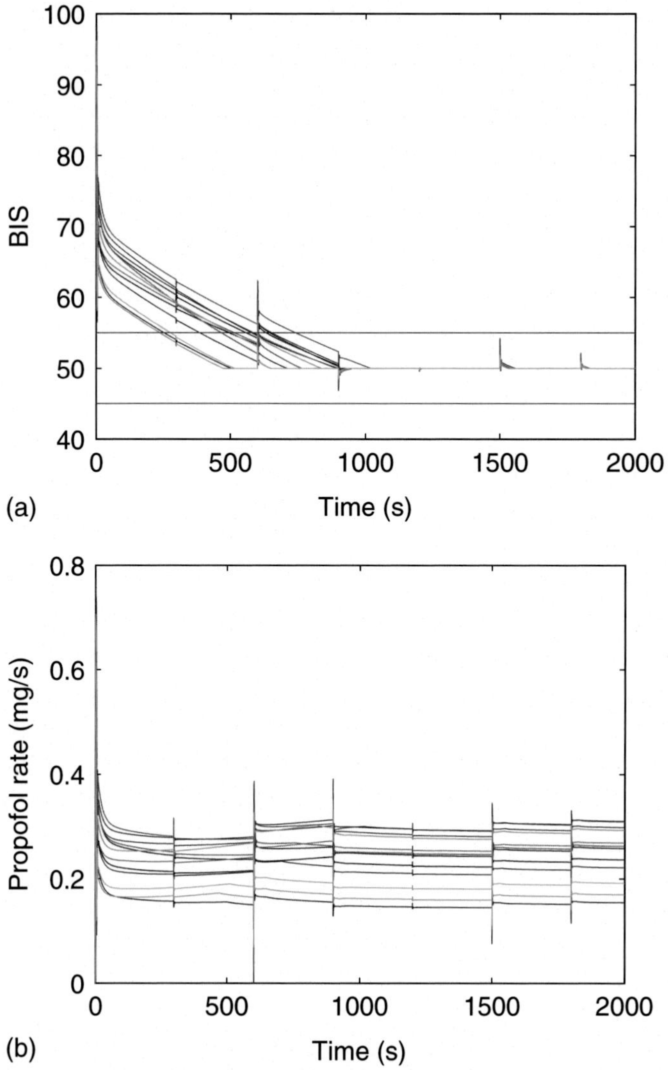

Figure 11.3 (a) Closed-loop response of first 2000 s of the output signal, BIS, during surgery for different patients; (b) closed-loop response of first 2000 s of the control signal, the rate of the injection of Propofol, during surgery for different patients.

It is further observed from the figures that the control signal during the maintenance phase never reaches its upper bound and hence it is a safe injection. Generally speaking, Propofol may cause some side effects and its injection rate should be kept as low as possible.

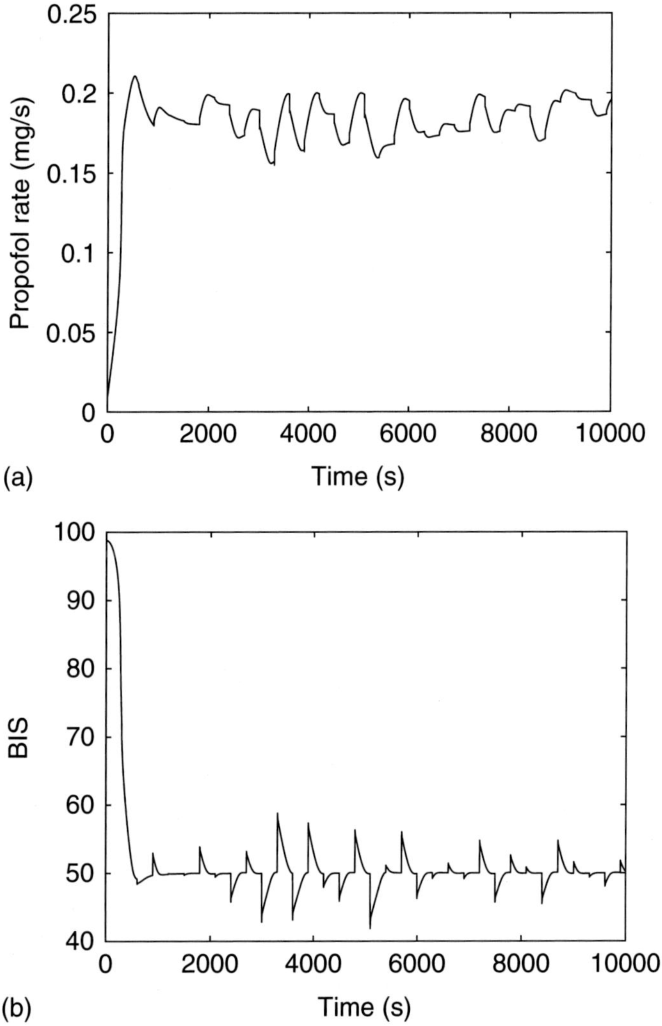

Figure 11.4 (a) Closed-loop response of the output variable, BIS, during surgery for one patient; (b) closed-loop response of the control signal, the rate of the injection of Propofol, during surgery for one patients.

11.2 CONTROL OF MAGNETIC RIGID SPACECRAFT

The dynamic behavior of a magnetic rigid spacecraft is known to be chaotic [10, 11], and its control has been considered in a number of works. The input-output feedback linearization approach is one of the methods that has been successfully applied to the control of a magnetic rigid spacecraft [12].

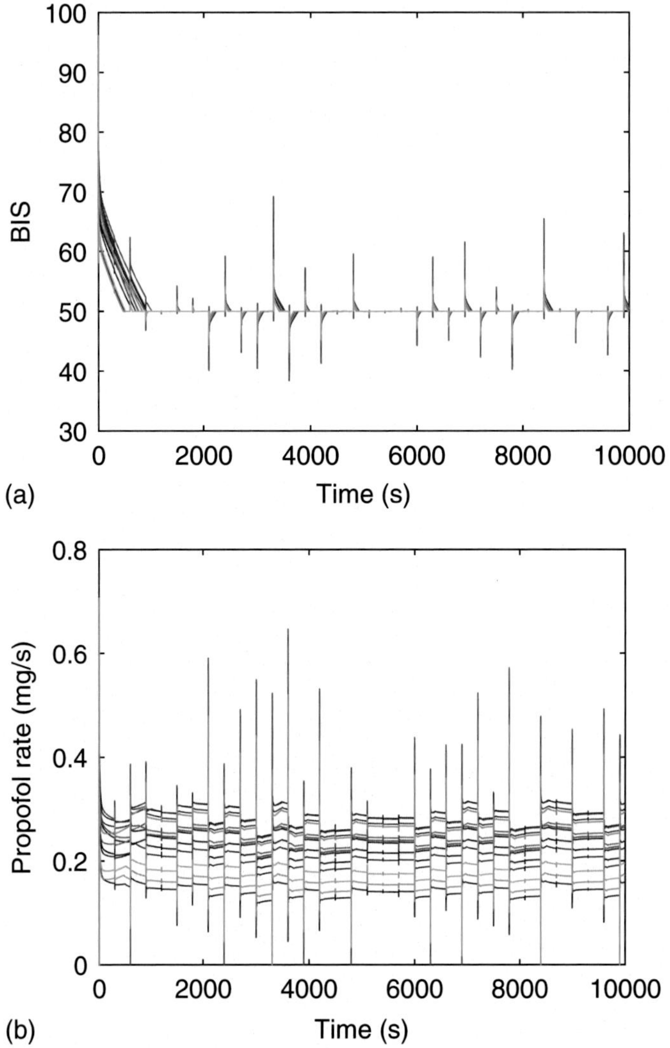

Figure 11.5 (a) Closed-loop response of the output signal, BIS, during surgery for different patients; (b) closed-loop response of the control signal, the rate of the injection of Propofol, during surgery for different patient.

In order to control the chaotic attitude motion of the spacecraft, a FEL control scheme, a T2FNN working in parallel with a PD controller, is used [19]. The T2FNN structure benefits from elliptic type-2 fuzzy MFs. The SMC theory-based parameter adaptation rules, proposed in Chapter 7, are used to tune the parameters of elliptic type-2 fuzzy MFs and the parameters of the consequent part.

11.2.1 Dynamic Model of a Magnetic Satellite

The magnetic rigid spacecraft is supposed to move in a circular orbit with the orbital angular velocity ω_c in the gravitational and the magnetic fields of the Earth. It is assumed that the inertial reference frame $(O_e - X_0Y_0Z_0)$ has the origin O_e at the mass center of the Earth with the polar axis of the Earth as Z_0-axis and the line from O_e to the ascending node as X_0-axis. Let $O - XYZ$ denote the orbital reference frame such that X is in the anti-nadir direction and Z is in the direction of the normal vector of the orbital plane XY. Let x_1 be the libration angle that represents the deviation of the spacecraft-fixed x and y axes from the orbital axes X and Y, respectively, the dynamics of the magnetic rigid spacecraft can then be expressed as (see, e.g., [10, 12]):

$$\begin{aligned}
\dot{x}_1 &= x_2 \\
\dot{x}_2 &= -K\sin(2x_1) - \gamma x_2 \\
&\quad - \alpha_s(2\sin(x_1)\sin(x_3) + \cos(x_1)\cos(x_3)) + u(t) \\
\dot{x}_3 &= 1
\end{aligned} \tag{11.9}$$

in which γ is the damping parameter, α_s is the magnetic parameter, t is the dimensionless time defined as the product of the orbital angular velocity and the actual time, and K is defined as:

$$K = \frac{3(B - A)}{2C} \tag{11.10}$$

where $A, B,$ and C are the principal inertia moments of the arbitrarily shaped spacecraft. The parameter K is equal to 1.1. Furthermore, the control signal u is $u = \frac{M_c}{C\omega_c^2}$, where M_c is the control torque. The dynamic behavior of the magnetic rigid spacecraft is known to show chaotic motion and its phase plane is highly sensitive to the values of γ and α_s and the initial conditions of the system. Figure 11.6 shows the phase plane of the zero-input system ($u = 0$) with different values for its parameter and initial conditions for the system. The system is simulated for 20 000 *sec* and the results are shown in Fig. 11.6, which shows that slight changes in the parameters of the spacecraft can greatly influence the trajectories of the system and the shape of the chaotic attractor.

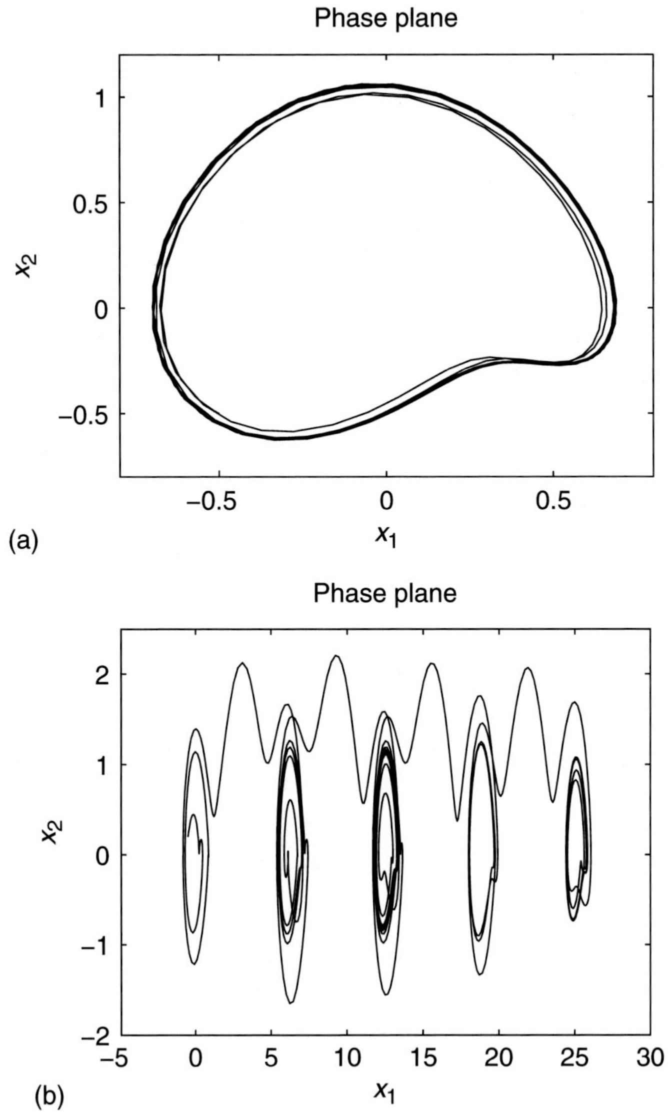

Figure 11.6 Phase portrait of magnetic rigid satellite with different values for the parameters and different initial states of the system. (a) The phase portrait with $\alpha_s = 0.69$, $K = 1.1$, $\gamma = 0.2$, $x_1 = -0.5$, $x_2 = -0.5$; (b) $\alpha_s = 0.7$, $K = 1.1$, $\gamma = 0.28$, $x_1 = -0.5$, $x_2 = 0.2$; *(Continued)*

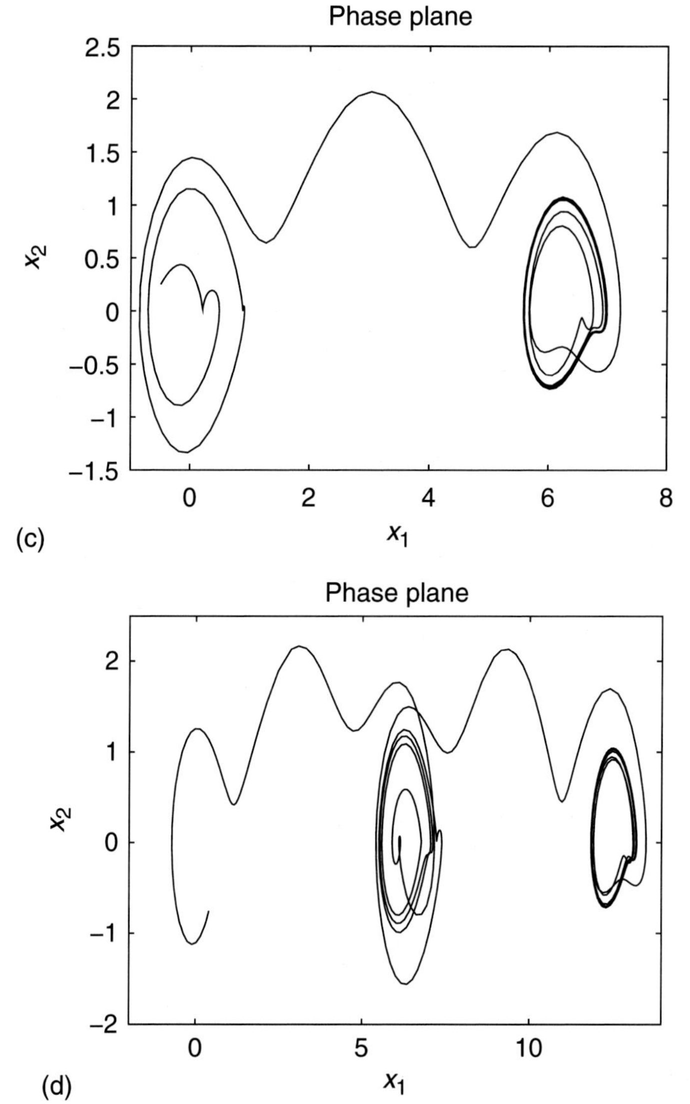

Figure 11.6, cont'd (c) $\alpha_s = 0.7, K = 1.1, \gamma = 0.285, x_1 = -0.5, x_2 = 0.25$; (d) $\alpha_s = 0.7, K = 1.1, \gamma = 0.297, x_1 = 0.4, x_2 = -0.75$.

11.2.2 Simulation Results

The proposed SMC-based learning algorithm for T2FNN is simulated on the attitude control of a magnetic satellite. The sample time for the simulation is selected as 0.001 *sec*. Furthermore, in order to make the

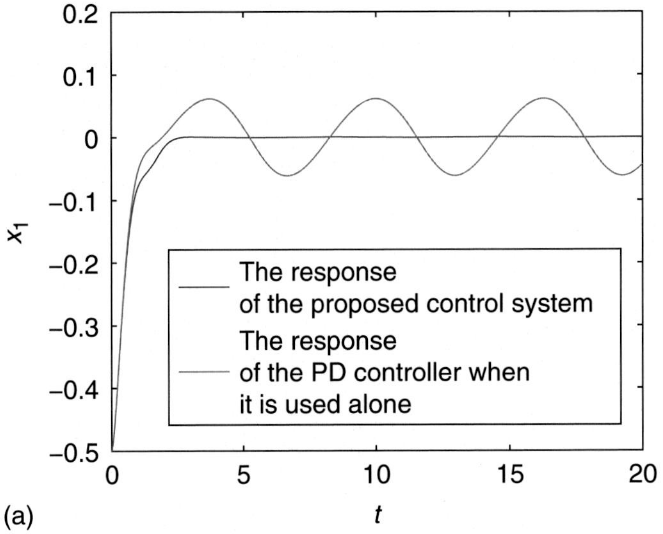

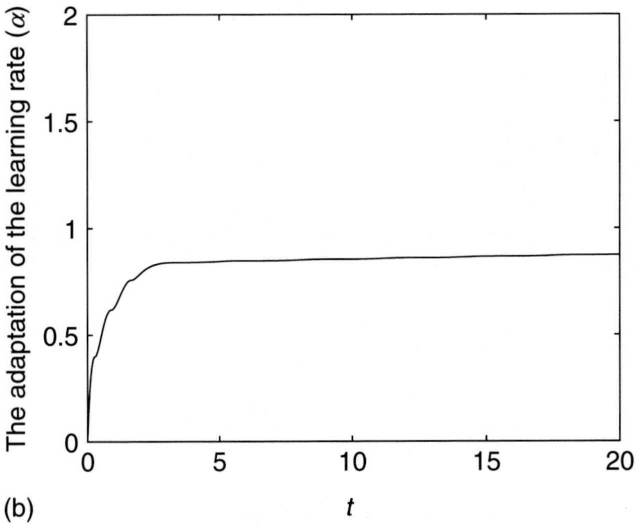

Figure 11.7 The regulation response of the magnetic satellite: (a) the libration angle in the orbital plane $x_1(t)$; (b) the evolution of the adaptive learning rate α; *(Continued)*

spacecraft behave chaotically its parameters are selected as α_s and γ, equal to 0.6984 and 0.2, respectively.

Figure 11.7 compares the regulation performance of the proposed FEL control structure and a PD controller working alone. The gains of the PD

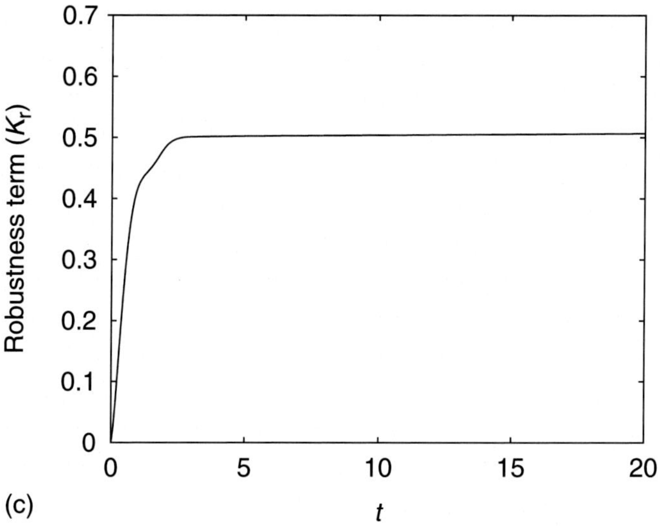

(c)

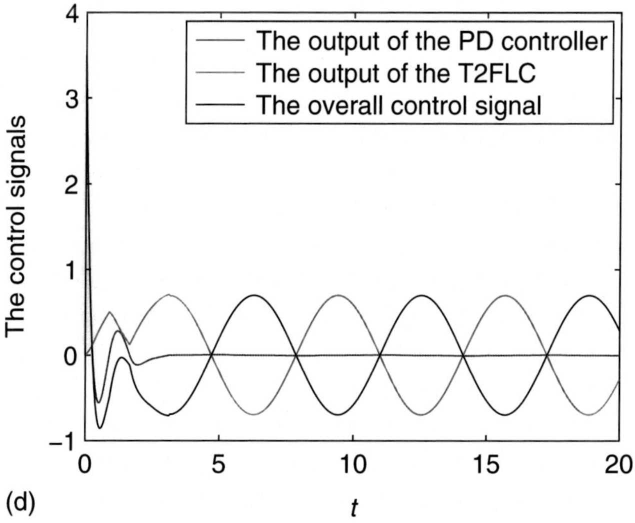

(d)

Figure 11.7, cont'd (c) the evolution of the robustness term K_r; (d) the control signals.

controller, K_P, and K_D are selected to be equal to 9 and 5, respectively. The initial conditions considered for the system are selected to be $x_1 = -0.5$ and $x_2 = 0.2$. As can be seen from Fig. 11.7, although the PD controller ensures the error signal is bounded in the neighborhood of zero, it cannot eliminate it. The interesting result is that although the system gives a steady-state error when only a PD controller is used, the fusion of the PD controller with a T2FNN eliminates the steady-state error. Moreover, the evolution of the robustness parameter K_r is shown in Fig. 11.7(c). As can be seen from Fig. 11.7, the initial value of K_r is selected to be equal to zero, and adaptation law as in Chapter 7 is used to find the optimal value of this parameter. In addition, Fig. 11.7(c) shows that the value of the parameter becomes as large as needed to ensure the robustness of the system. The evolution of the adaptive learning rate α is shown in Fig. 11.7(b). The adaptation law for the learning rate α makes it possible to control the system without any a priori knowledge about the upper bound of the states of the system. Figure 11.7(d) shows the overall control signal (τ), the output of T2FNN (τ_f), and the output of the conventional PD controller (τ_c). As can be seen from the figure, at the beginning of the simulation, the overall control signal is mostly due to the conventional PD controller. After finite time, T2FNN learns the dynamics of the system and takes responsibility for the system. Simultaneously, the output of the PD controller tends to go to zero.

Figure 11.8 compares the tracking performance of the proposed control approach and a PD controller when it is used alone. Furthermore, it can be seen from this figure that the PD controller makes the error signal bounded in a neighborhood near zero but it cannot make it zero. Figure 11.8(b) shows the evolution of the learning rate (α) over time. The use of the adaptation law for the learning rate makes it possible to adjust it during training, and no more trial and error is needed to tune this parameter. The evolution of the parameter K_r is depicted in Fig. 11.8(c). The initial value of K_r is zero, and the adaptation law as in Chapter 7 is used to tune this parameter. Figure 11.8(d) shows the overall control signal (τ), the output of T2FNN (τ_f), and the output of the conventional PD controller (τ_c).

Figure 11.9(a) compares the tracking performance of the proposed algorithm with that of the PD controller when the reference signal is sinusoidal. As can be seen from Fig. 11.9, the proposed FEL structure also outperforms the PD controller for the case of the sinusoidal reference input, and there is no steady-state error in the system.

T2FLSs have more degrees of freedom to deal with noise and are a better choice when there is a high level of noise in the system. In order to compare the performance of T2FNN with its type-1 counterpart under noisy conditions, uniformly distributed white noise with a standard deviation of 0.03 is added to the measurement. Figure 11.10 shows the percentage of improvement of T2FLS over a type-1 counterpart for different

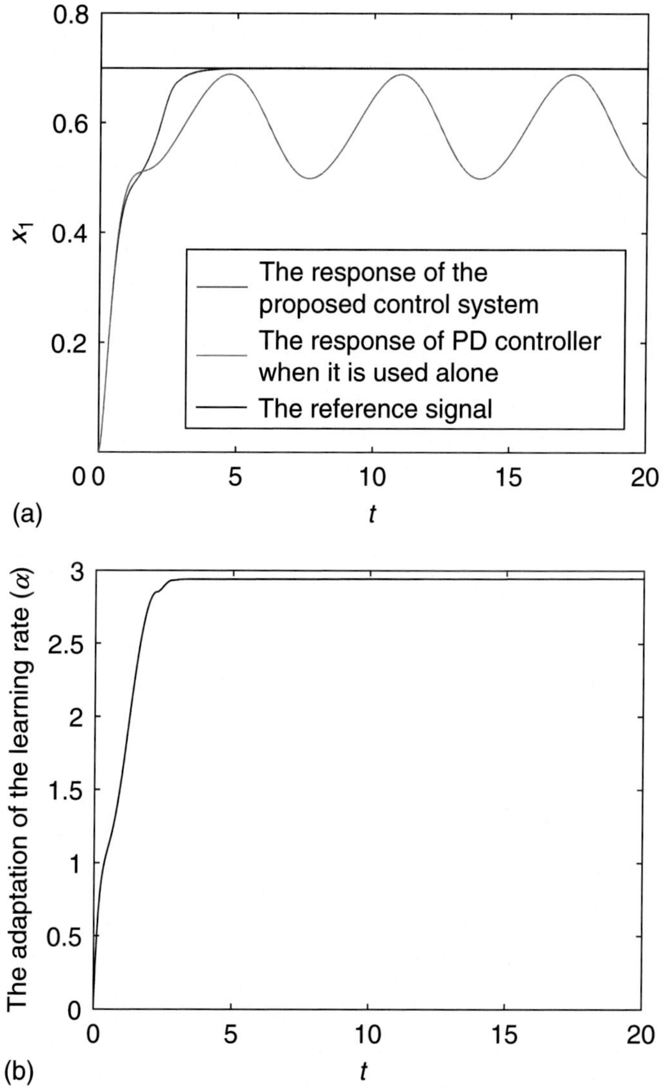

Figure 11.8 The tracking response of the magnetic satellite: (a) the libration angle in the orbital plane $x_1(t)$; (b) the evolution of the adaptive learning rate α; *(Continued)*

initial values of the a_1 and a_2 parameters of T2FLS. As can be seen from Fig. 11.10, it is possible to select such initial values for a_1 and a_2 that the system is controlled with 40% less integral of squared error than T1FLS. The figure also shows that an appropriate interval for the selection of the initial

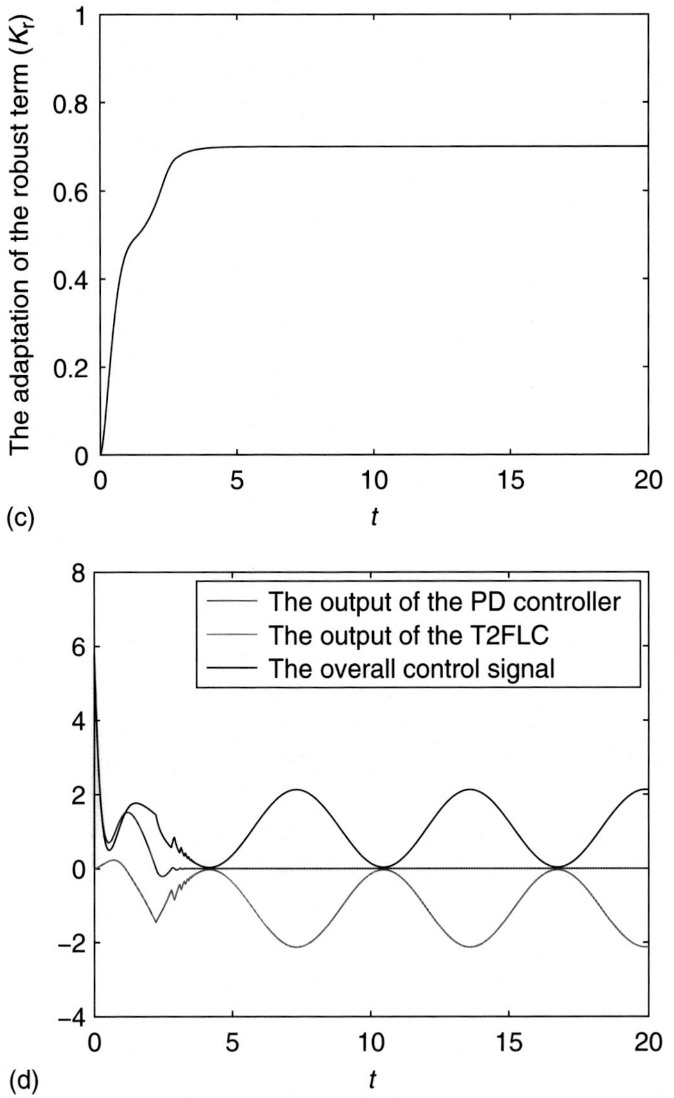

Figure 11.8, cont'd (c) the evolution of the robustness term K_r; (d) the control signals.

values a_1's and a_2's is $1.65 < a_1 < 1.85$ and $0.75 < a_2 < 1$. These findings comply with those of other works seen in the literature [13–17], suggesting that the use of type-2 fuzzy systems adds more degrees of freedom to the design. The end result is that it is possible to cope with noisy measurements and uncertainties in the system more effectively.

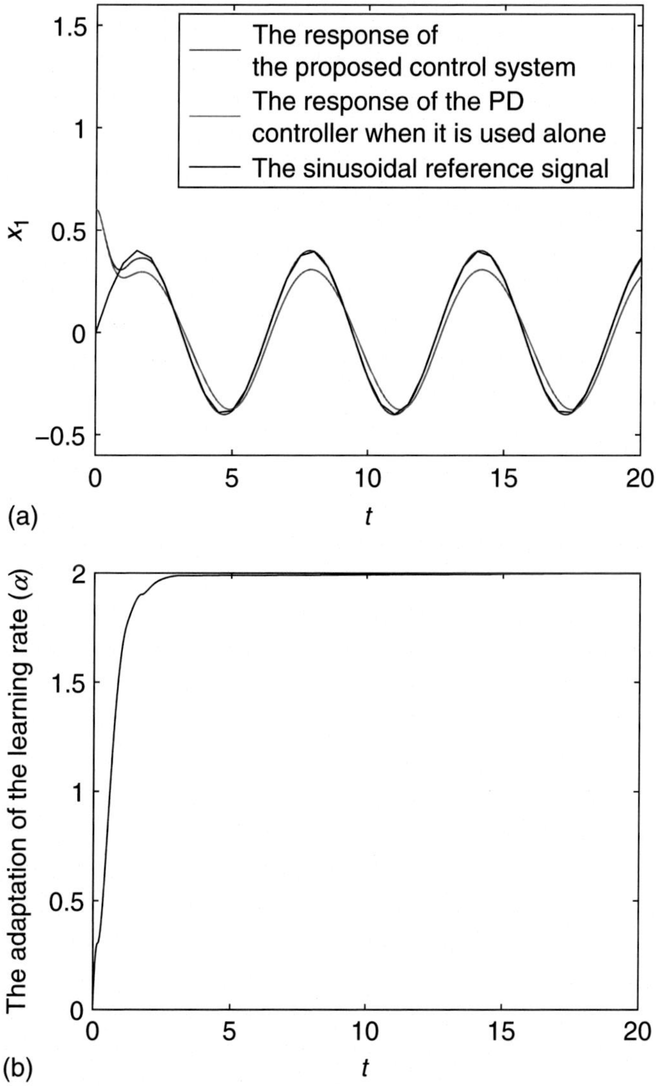

Figure 11.9 Tracking response of the magnetic satellite when the reference signal is sinusoidal: (a) the libration angle in the orbital plane $x_1(t)$; (b) the evolution of the adaptive learning rate α; *(Continued)*

11.3 CONTROL OF AUTONOMOUS TRACTOR

Autonomous navigation of an agricultural vehicle involves the control of different dynamic subsystems, such as the yaw angle dynamics and the longitudinal speed dynamics. In this section, a PID controller is used to

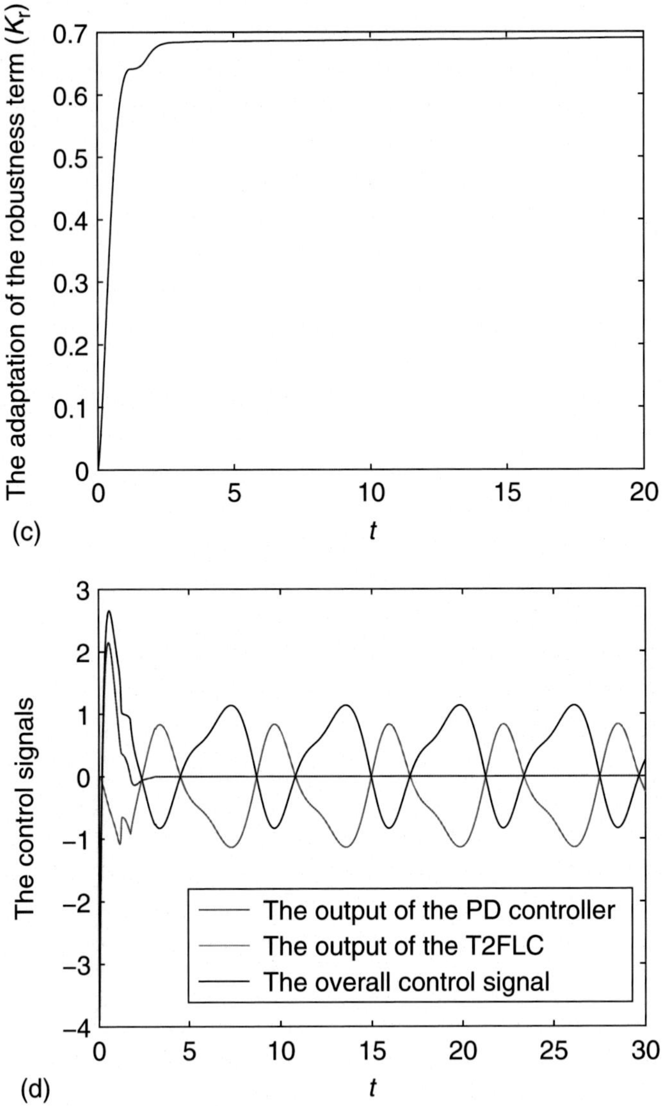

Figure 11.9, cont'd (c) the evolution of the robustness term K_r; (d) the control signals.

control the longitudinal velocity of the tractor. For the control of the yaw angle dynamics, a PD controller works in parallel with a T2FNN. In this way, instead of modeling the interactions between the subsystems prior to the design of a model-based control, we develop a control algorithm that learns the interactions online from the measured feedback

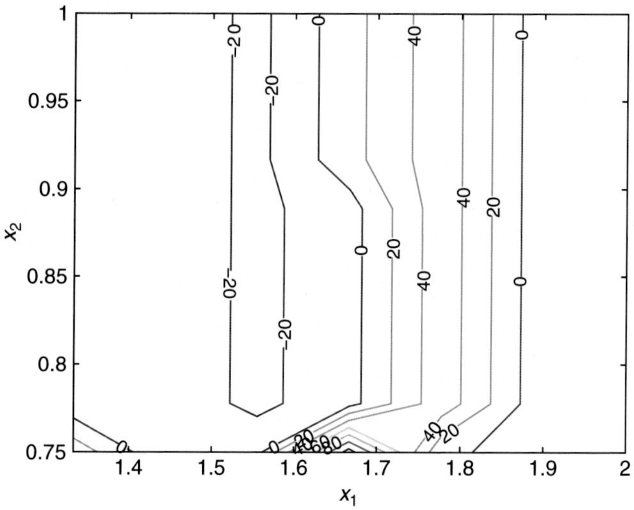

Figure 11.10 Percentage of improvement of type-2 fuzzy system over type-1 counterpart in noisy conditions.

error. In addition to the control of the stated subsystems, a kinematic controller is needed to correct the errors in both the x and y axis for the trajectory tracking problem of the tractor. To demonstrate the real-time abilities of the proposed control scheme, an autonomous tractor is equipped with the use of reasonably priced sensors and actuators. Experimental results show the efficacy and efficiency of the proposed learning algorithm.

The motivation behind the use of self-learning controllers instead of conventional controllers for the control of agricultural production machines is that there are different subsystems interacting with each other in these machines, and well tuning of the controller coefficients simultaneously is a difficult task. Even if the operator becomes proficient in proper adjustment of the different controller coefficients, crop/animal variability and environmental conditions force the operator to change the machine settings continuously.

11.3.1 Mathematical Description of Tractor

11.3.1.1 Kinematic Model

A schematic diagram of the autonomous tractor is presented in Fig. 11.11.

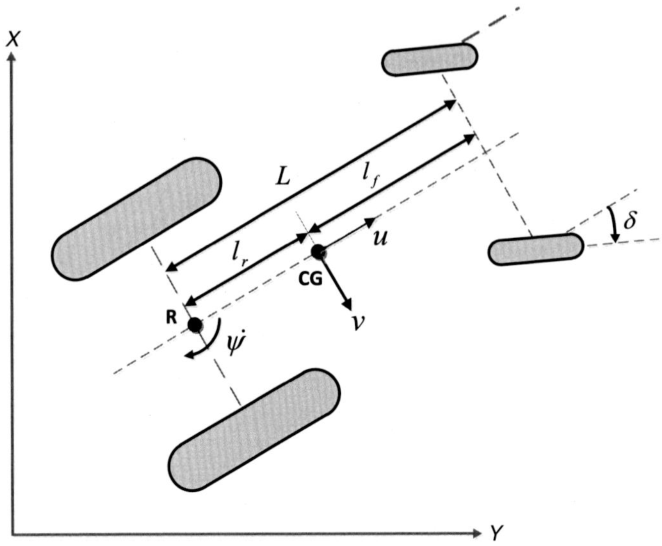

Figure 11.11 Autonomous tractor.

The linear velocities $\dot{x}$, $\dot{y}$ and the yaw rate $\dot{\psi}$ at the R point are written as follows [20]:

$$\begin{bmatrix} \dot{x} \\ \dot{y} \\ \dot{\psi} \end{bmatrix} = \begin{bmatrix} u\cos\psi \\ u\sin\psi \\ \frac{u\tan\delta}{L} \end{bmatrix} \tag{11.11}$$

where u, ψ, δ, and L represent the longitudinal velocity, the yaw angle defined on the point R on the tractor, the steering angle of the front wheel, the distance between the front, and the rear axles of the tractor, respectively.

Considering the center of gravity shown in Fig. 11.11, the linear velocities $\dot{x}$, $\dot{y}$ and the yaw rate $\dot{\psi}$ of center of gravity can be written as follows [20]:

$$\begin{bmatrix} \dot{x} \\ \dot{y} \\ \dot{\psi} \end{bmatrix} = \begin{bmatrix} u\cos\psi - v\sin\psi \\ u\sin\psi + v\cos\psi \\ \frac{u\tan\delta}{L} \end{bmatrix} \tag{11.12}$$

where v equals the multiplication of $\dot{\psi}$ and l_r.

11.3.1.2 Yaw Dynamics Model

The velocities and side-slip angles on the rigid body of the tractor are presented in Fig. 11.12(a). Similarly, the forces on the rigid body of the tractor are shown in Fig. 11.12(b).

The equations of yaw motion of the autonomous tractor are written in state-space form as follows [20]:

$$\begin{bmatrix} \dot{v} \\ \dot{\gamma} \end{bmatrix} = \begin{bmatrix} A_{11} & A_{12} \\ A_{21} & A_{22} \end{bmatrix} \begin{bmatrix} v \\ \gamma \end{bmatrix} + \begin{bmatrix} B_1 \\ B_2 \end{bmatrix} \delta(t) \tag{11.13}$$

where:

$$\begin{aligned} A_{11} &= -\frac{C_{\alpha,f} + C_{\alpha,r}}{mu}, \\ A_{12} &= \frac{-l_f C_{\alpha,f} + l_r C_{\alpha,r}}{mu} - u, \\ A_{21} &= \frac{-l_f C_{\alpha,f} + l_r C_{\alpha,r}}{I_z u}, \\ A_{22} &= -\frac{l_f^2 C_{\alpha,f} + l_r^2 C_{\alpha,r}}{I_z u}, \\ B_1 &= \frac{C_{\alpha,f}}{m}, \quad B_2 = \frac{l_f C_{\alpha,f}}{I_z} \end{aligned} \tag{11.14}$$

where m, v, u, γ, and δ represent the mass of the tractor, the lateral velocity of the center of gravity, the longitudinal velocity of the center of gravity, the yaw rate, and the steering angle of the front wheel, respectively. Furthermore, l_f, l_r and I_z, respectively, represent the distance between the front axle and the center of gravity of the tractor, the distance between the rear axle, the center of gravity of the tractor, and the inertial moment of the tractor. The term $C_{\alpha,i}$ represents the cornering stiffness of the tires of the tractor.

11.3.2 Overall Control Scheme

11.3.2.1 Kinematic Controller

The kinematic model is rewritten in state-space as follows:

$$\begin{bmatrix} \dot{x} \\ \dot{y} \\ \dot{\psi} \end{bmatrix} = \begin{bmatrix} \cos\psi & -l_r \sin\psi \\ \sin\psi & l_r \cos\psi \\ 0 & 1 \end{bmatrix} \begin{bmatrix} u \\ \gamma \end{bmatrix} \tag{11.15}$$

where lateral velocity v equals γl_r.

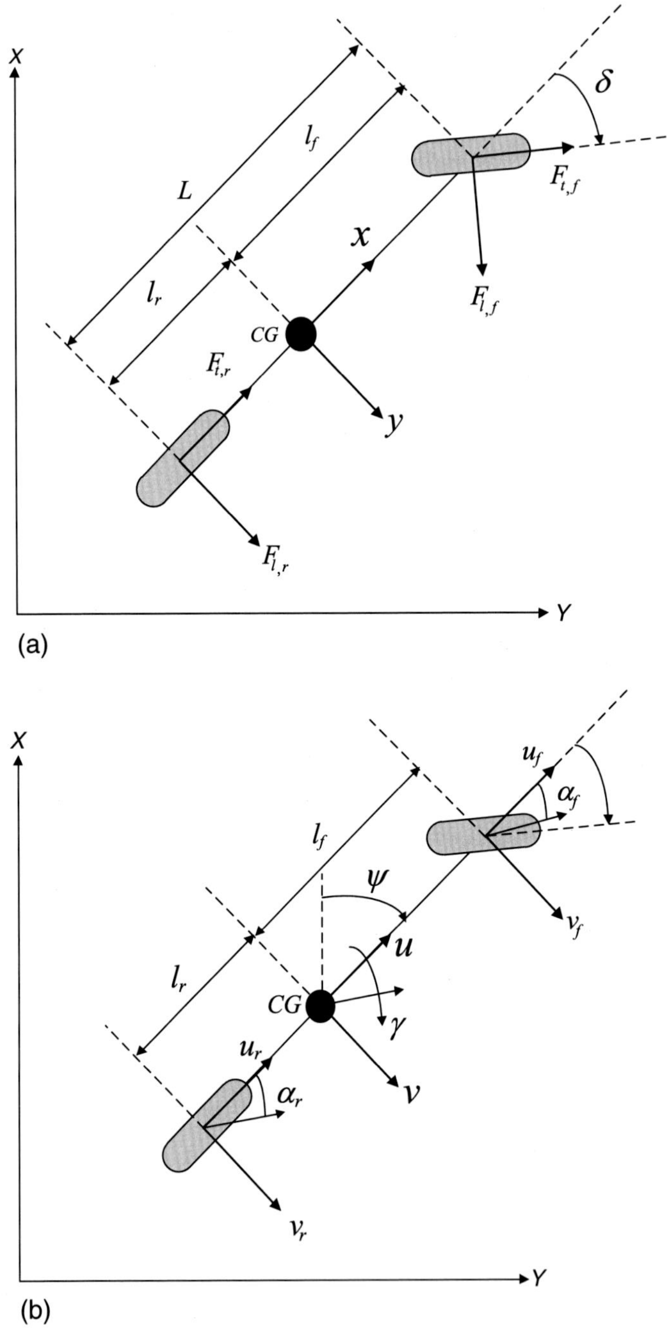

Figure 11.12 Dynamic bicycle model for a tractor: (a) velocities and side-slip angles; (b) forces on the rigid body of the system.

An inverse kinematic model is needed to calculate the reference speed and the yaw rate for the tractor. It is written as follows:

$$\begin{bmatrix} u \\ \gamma \end{bmatrix} = \begin{bmatrix} \cos\psi & \sin\psi \\ -\frac{1}{l_r}\sin\psi & \frac{1}{l_r}\cos\psi \end{bmatrix} \begin{bmatrix} \dot{x} \\ \dot{y} \end{bmatrix} \tag{11.16}$$

The kinematic control law proposed in Ref. [18] to be applied to the tractor for trajectory tracking control is written as:

$$\begin{bmatrix} u_{ref} \\ \gamma_{ref} \end{bmatrix} = \begin{bmatrix} \cos\psi & \sin\psi \\ -\frac{1}{l_r}\sin\psi & \frac{1}{l_r}\cos\psi \end{bmatrix} \begin{bmatrix} \dot{x}_d + k_s \tanh(k_e e_x) \\ \dot{y}_d + k_s \tanh(k_e e_y) \end{bmatrix} \tag{11.17}$$

where $e_x = x_d - x$ and $e_y = y_d - y$ are the current position errors in the axes X and Y, respectively. The parameter k_e is the gain of the controller and k_s is the saturation constant. The coordinates (x, y) and (x_d, y_d) are the current and the desired coordinates, respectively. The parameters u_{ref} and γ_{ref} are the generated references for the speed and the yaw rate controllers.

11.3.2.2 Dynamic Controllers

The proposed control scheme used in this study is illustrated in Fig. 11.13. The arrow in Fig. 11.13 indicates that the output of the PD controller is used to tune the parameters of the T2FNN. The output of the PD+T2FNN controller is the steering angle of the front wheel. A low-level controller is used to control the steering mechanism. A PID controller is used for the control of longitudinal velocity. For the control of yaw dynamics, a conventional controller (such as a PD controller) works in parallel with an intelligent controller. On the other hand, the longitudinal dynamics could, of course, be selected. Moreover, there could be two T2FNNs running on the control of the two subsystems (yaw dynamics and longitudinal dynamics) simultaneously. The reason for such a selection in this study is that the yaw dynamics control of the tractor is more common in agricultural machines. Even if a time-based trajectory (both the yaw angle and the longitudinal speed of the tractor are controlled simultaneously) is given to the system in this study, a space-based trajectory (the longitudinal speed is fixed and only the yaw dynamics is controlled) is also very common in agricultural applications. Based on these concerns, the yaw dynamics of the tractor have been chosen for the implementation of the novel learning algorithm proposed here.

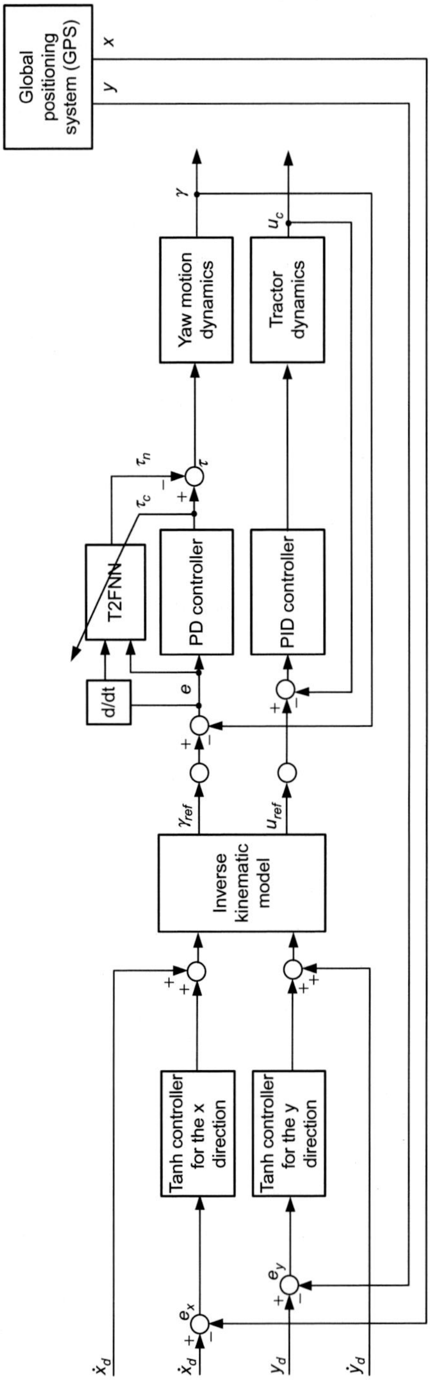

Figure 11.13 Block diagram of the proposed adaptive fuzzy neuro scheme.

11.3.3 Experimental Results

An 8-shaped reference trajectory is applied to the system. The reference and the actual trajectories of the system, both the longitudinal and the lateral error values on the related trajectory, are shown in Fig. 11.14(a),

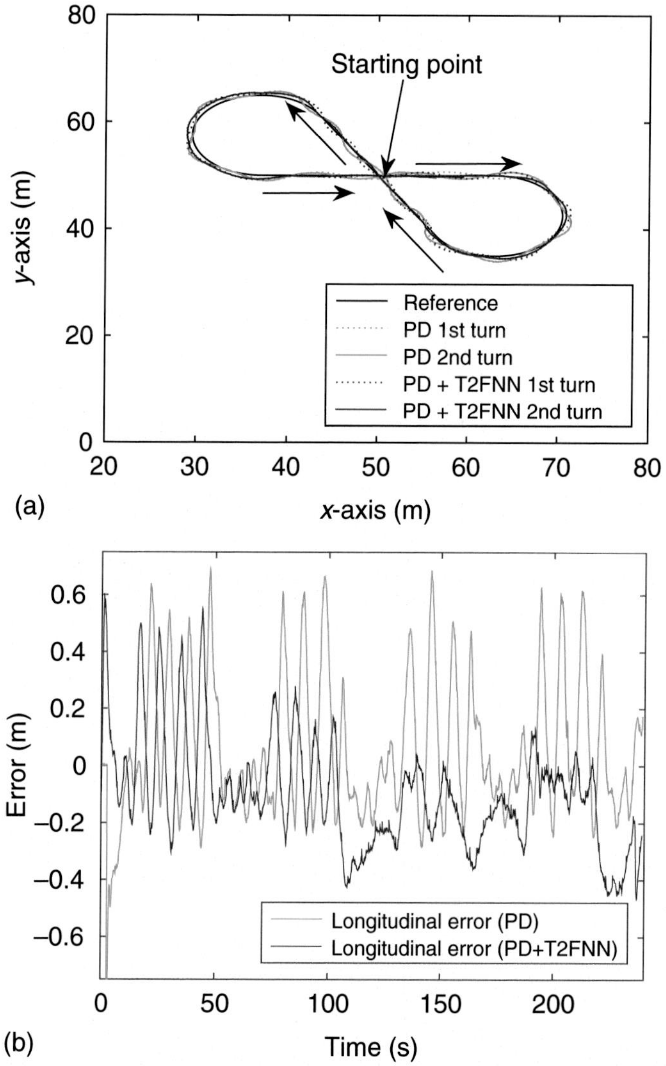

Figure 11.14 (a) The trajectory of the autonomous tractor; (b) the longitudinal error; *(Continued)*

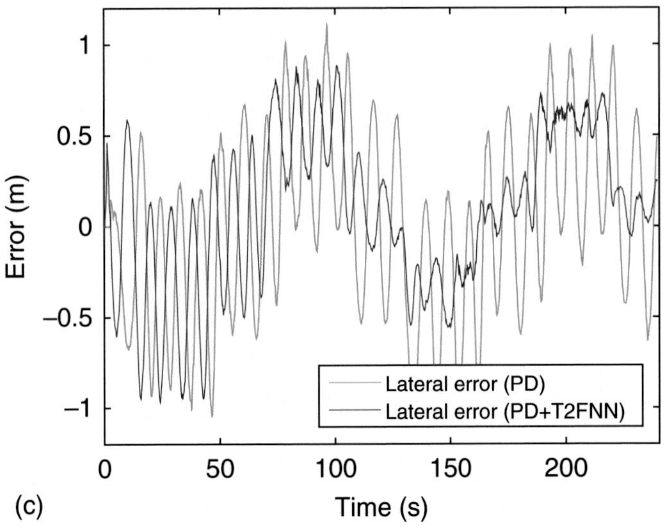

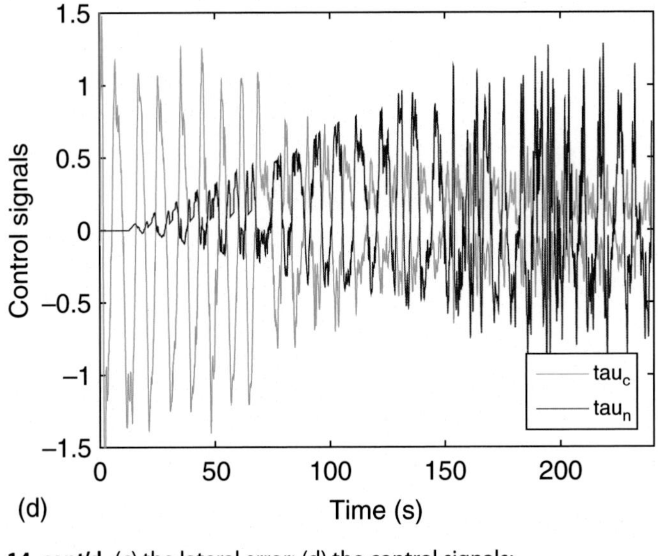

Figure 11.14, cont'd (c) the lateral error; (d) the control signals;

(b), and (c) for two different controllers, respectively. The results show that the control scheme consisting of a T2FNN working in parallel with a PD controller gives a better trajectory following accuracy than the one where only a PD controller acts alone. It can be argued that the performance of the conventional controller acting alone can be improved by better tuning,

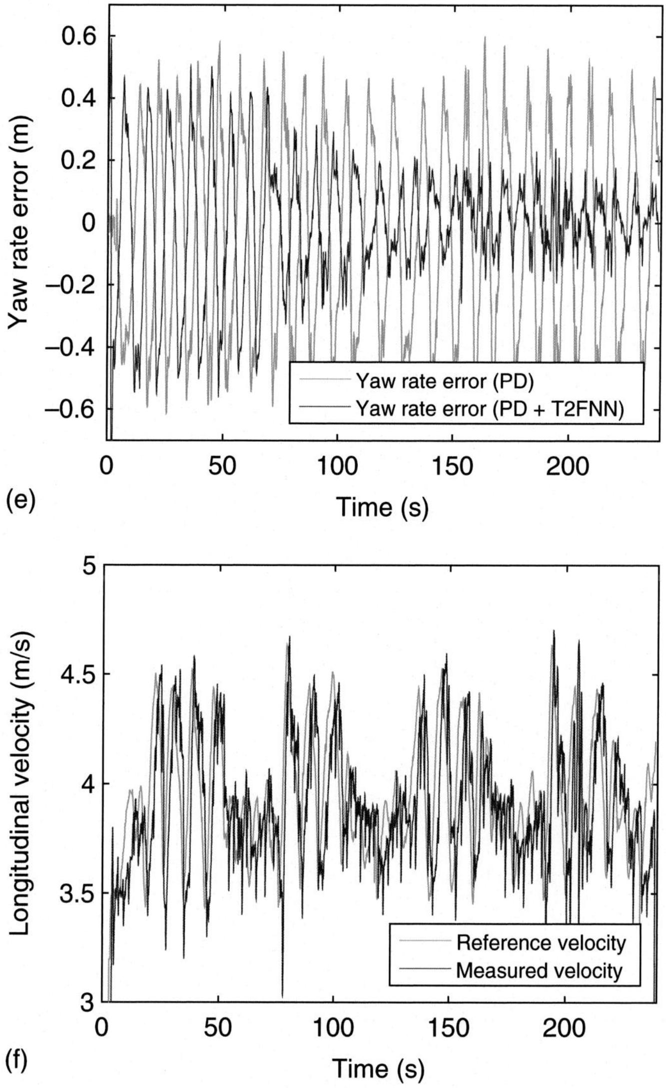

Figure 11.14, cont'd (e) the yaw rate error; (f) actual longitudinal velocity and measured longitudinal velocity.

but as has already been stated, in real life this is a challenging task because in addition to the interactions of the subsystems, there exist unmodeled dynamics and uncertainties in real-world applications. Thus, the proposed control structure, consisting of an intelligent controller and a conventional controller, would be preferable in real life.

In Fig. 11.14(a), while the dotted lines show the first turns, the solid lines represent the second turns. The control accuracy for the PD controller is the same for the first and second turn, which is expected. However, the control accuracy for the T2FNN working in parallel with a PD controller is better for the second turn. We observe similar behavior in Fig. 11.14(b) and (c). The results in Fig. 11.14(c) shows performance improvement of 30% in the case of a PD controller working in parallel with the T2FNN.

Figure 11.14(d) shows the control signals coming from the conventional PD controller and T2FNN when the PD controller works in parallel with the T2FNN. As can be seen from Fig. 11.14(d), at the beginning (in the first turn), the dominating control signal is the one coming from the PD controller. After the first turn (starting from the 120th second), the T2FNN is able to take over the control, thus becoming the leading controller. Moreover, when the reference signal changes, the output of the PD controller increases. However, after a finite time, the output of the PD controller again comes back to approximately zero.

Although there exist two independent subsystem controllers in the autonomous tractor control system, the T2FNN works in parallel with a PD controller only for the control of the yaw rate of the system. Thus, the error signal for both PD controller acting alone and in parallel with the T2FNN are shown in Fig. 11.14(e). As can be seen from Fig. 11.14(e), the T2FNN significantly increases the control accuracy of the yaw dynamics of the system, since the followed trajectory is a time base trajectory.

Further, since the reference trajectory to be followed in this experiment is not a spaced-based trajectory, but a time-based one, the longitudinal velocity control performance has a direct and significant effect on the accuracy of the trajectory tracking. For tracking the longitudinal velocity reference, a structure consisting of a PID controller and a feed-forward controller is used. As can be seen Fig. 11.14(f), the control performance of the longitudinal controller is satisfactory.

11.4 CONCLUSION

The simulation results in all these three control case studies show that when the system is controlled by using a conventional controller in parallel with a T2FNN, the accuracy of the overall controller increases. In this method, the conventional controller is responsible for the stability of the system at the beginning of the learning process. After the learning process starts, the T2FNN controller learns the system dynamics and takes responsibility for controlling the system gradually. In other words, there is no need to well-tune the conventional controller. Another advantage of the proposed scheme is that the SMC theory-based parameters update rules for the T2FNN proposed are simpler compared to other methods such as gradient descent, because the proposed parameter adaptation rules have neither matrix manipulations nor partial derivatives.

REFERENCES

[1] R. Hovorka, Closed-loop insulin delivery: from bench to clinical practice, Nature Rev. Endocrinol. 7 (7) (2011) 385-395.
[2] J. Reid, G. Kenny, Evaluation of closed-loop control of arterial pressure after cardiopulmonary bypass, Brit. J. Anaest. 59 (2) (1987) 247-255.
[3] H. Puebla, J. lvarez Ramírez, A cascade feedback control approach for hypnosis, Ann. Biomed. Eng. 33 (10) (2005) 1449-1463, ISSN 0090-6964.
[4] C.M. Ionescu, R. De Keyser, B.C. Torrico, T. De Smet, M. Struys, J.E. Normey-Rico, Robust predictive control strategy applied for propofol dosing using BIS as a controlled variable during anesthesia, IEEE Trans. Biomed. Eng. 55 (9) (2008) 2161-2170.
[5] M. El-Bardini, A.M. El-Nagar, Direct adaptive interval type-2 fuzzy logic controller for the multivariable anaesthesia system, Ain Shams Eng. J. 2 (3) (2011) 149-160.
[6] T. Bouillon, J. Bruhn, L. Radu-Radulescu, E. Bertaccini, S. Park, S. Shafer, Non-steady state analysis of the pharmacokinetic interaction between propofol and remifentanil, Anesthesiology 97 (6) (2002) 1350-1362.
[7] C.S. Nunes, T. Mendonca, S. Bras, D.A. Ferreira, P. Amorim, Modeling anesthetic drugs' pharmacodynamic interaction on the bispectral index of the EEG: the influence of heart rate, in: 29th Annual International Conference of the IEEE on Engineering in Medicine and Biology Society (EMBS 2007), IEEE, 2007, pp. 6479-6482.
[8] C. Lennmarken, M.-L. Lindholm, S.D. Greenwald, R. Sandin, Confirmation that low intraoperative BIS^{TM} levels predict increased risk of post-operative mortality, Anesthesiology 99 (3) (2003) A30.
[9] T. Zikov, S. Bibian, G.A. Dumont, M. Huzmezan, C.R. Ries, A wavelet based de-noising technique for ocular artifact correction of the electroencephalogram, in: Engineering in Medicine and Biology, 2002, Proceedings of the Second Joint 24th Annual Conference and the Annual Fall Meeting of the Biomedical Engineering Society EMBS/BMES Conference, 2002, vol. 1, IEEE, 2002, pp. 98-105.
[10] Y. LIU, L. CHEN, Chaos in attitude dynamics of spacecraft, Springer, New York, 2013.
[11] L.-Q. Chen, Y.-Z. Liu, G. Cheng, Chaotic attitude motion of a magnetic rigid spacecraft in a circular orbit near the equatorial plane, J. Fran. Inst. 339 (1) (2002) 121-128.
[12] L.-Q. Chen, Y.-Z. Liu, Chaotic attitude motion of a magnetic rigid spacecraft and its control, Int. J. Non-linear Mech. 37 (3) (2002) 493-504.
[13] R. Sepulveda, P. Melin, A. Rodriguez, A. Mancilla, O. Montiel, Analyzing the effects of the footprint of uncertainty in type-2 fuzzy logic controllers, Eng. Lett. 13 (2006) 138-147.
[14] J.M. Mendel, Uncertainty, fuzzy logic, and signal processing, Signal Process. 80 (2000) 913-933.

[15] M.-Y. Hsiao, T.-H.S. Li, J.Z. Lee, C.H. Chao, S.H. Tsai, Design of interval type-2 fuzzy sliding-mode controller, bibinfojournalInformat. Sci. Int. J. 178 (2008) 1696-1716.
[16] M.A. Khanesar, E. Kayacan, M. Teshnehlab, O. Kaynak, Analysis of the noise reduction property of type-2 fuzzy logic systems using a novel type-2 membership function, IEEE Trans. Syst. Man Cybernet. B Cybernet. 41 (5) (2011) 1395-1406.
[17] M.A. Khanesar, E. Kayacan, M. Teshnehlab, O. Kaynak, Extended Kalman filter based learning algorithm for type-2 fuzzy logic systems and its experimental evaluation, IEEE Indust. Electron. 59 (11) (2012) 4443-4455.
[18] F.N. Martins, W.C. Celeste, R. Carelli, M. Sarcinelli-Filho, T.F. Bastos-Filho, An adaptive dynamic controller for autonomous mobile robot trajectory tracking, Control Eng. Pract. 16 (11) (2008) 1354-1363.
[19] M.A. Khanesar, E. Kayacan, M. Reyhanoglu, O. Kaynak, Feedback Error Learning Control of Magnetic Satellites Using Type-2 Fuzzy Neural Networks With Elliptic Membership Functions, Cybernetics, IEEE Transactions on 45 (4) (2015) 858-868.
[20] E. Kayacan, E. Kayacan, H. Ramon, O. Kaynak, W. Saeys, Towards Agrobots: Trajectory Control of an Autonomous Tractor Using Type-2 Fuzzy Logic Controllers, IEEE/ASME Trans. Mechatron. 20 (1) (2015) 287-298, ISSN 1083-4435, doi:10.1109/TMECH.2013.2291874.

Appendix A

Contents

Abstract

This chapter provides sample source code for number of training algorithms for T2FNNs. The programming language is Matlab.

Keywords

Simulations, Mackey-Glass, Matlab source code, T2FNN Training using Matlab

A.1 SMC THEORY-BASED TRAINING ALGORITHM FOR TYPE-2 FUZZY NEURAL NETWORK

The main script file for the full SMC theory-based training algorithm for T2FNN is as follows. In order to execute this script, several functions are needed. These functions are "t2fnnsmc.m," "MatrixMultiple.m," "scale.m'," and "inv_scale.m." The source code for these functions must be written independently. This source code is provided after the main script.

A.1.1 Main Script for SMC Theory-Based Training Algorithm for T2FNN

```
clear all
close all
clc

```

```
%%%%%%%%%%%%%%%%%%%%%%%
%%% Data generation %%%
%%%%%%%%%%%%%%%%%%%%%%%
x(1:18)=1.2; %Initial conditions for Mackey-Glass
for t=18:2000
   x(t+1)=x(t)-0.1*x(t)+0.2*x(t-17)/(x(t-17)^10+1);
   %The dynamic of mackey-Glass
end
x=x(18:end);

%%%%%%%%%%%%%%%%%%%%%%%%%%
%%% Normalization part %%%
%%%%%%%%%%%%%%%%%%%%%%%%%%
minn=0; maxn=1;
norm_x=minmax(x); y=scale(x,norm_x(1),norm_x(2),minn,maxn);
% Target normalization
%%%%%%%%%%%%%%%%%%%%%%%%%%%%%%%%%
%%% Data prepration for T2FNN %%%
%%%%%%%%%%%%%%%%%%%%%%%%%%%%%%%%%
tbeg=1; tend=length(y);
for t=2:1501
     Data(t,:)=[y(t) y(t+6) y(t+12) y(t+18) y(t+24)];
     % [y(t-18) y(t-12) y(t-6) y(t)] is the input and
     % y(t+6) is the target value of the T2FNN
     Data_old(t-1,:)=[y(t-1) y(t+5) y(t+11) y(t+17) y(t+23)];
end
[row,col]=size(Data);

%%%%%%%%%%%%%%%%%%%%%%%%%%%%
%%% T2FNN initialization %%%
%%%%%%%%%%%%%%%%%%%%%%%%%%%%
q=0.5; % The initial condition for the q
alpha=0; % The initial condition for the learning rate of the
consequent part
center=[0.25 0.75]'; % The initial values of the centers of
MFs for T2FNN
for i=1:col-1
     C(:,i)=center;
end
S_upp=0.3*ones(length(center),col-1); % The initial values of
% the sigmas
S_low=0.2*ones(length(center),col-1); % The initial values of
% the sigmas
ar=rand(col-1,length(center)^(col-1));
% The initial values of the gains of inputs in the consequent
% part
b=rand(1,length(center)^(col-1))*1e-2;
```

```
% The initial values of the constant terms in the consequent
% part

delta_t=0.0001; %The sample time

%%%%%%%%%%%%%%%%
%%% Main loop %%%
%%% Training %%%
%%%%%%%%%%%%%%%%
for epoch=1:100
    for i=3:1000

        [T2FNN_output(i),C,S_low,S_upp,ar,b,q,alpha] =
        t2fnnsmc(Data(i,:),...
        Data_old(i,:),C,S_low,S_upp,ar,b,delta_t,q,alpha);
            % The calculation of the output of T2FNN and its
            % update procedure
            % based on SMC-Theory based training algorithm

        e(i-2)=T2FNN_output(i)-Data(i,col); %The training error
   end
   qq(epoch)=q; % The updated value of q
   learning_rate(epoch)=alpha;
   % The value of the learning rate obtained at the end of
   % each epoch
   Modeling_Error(epoch)=sqrt(mse(e));
   % The value of the RMSE obtained at the end of each epoch
end


%%%%%%%%%%%%%%%%
%%% Testing %%%
%%%%%%%%%%%%%%%%
for i=1001:1200
    T2FNN_output(i) = t2fnnsmc(Data(i,:),Data_old(i,:),...
       C,S_low,S_upp,ar,b,delta_t,q,alpha); %Output
       % evaluation

    et(i-1000)=T2FNN_output(i)-Data(i,col);
end

%%%%%%%%%%%%%%%%
%%% Display %%%
%%%%%%%%%%%%%%%%
display(['The RMSE of the training is equal to '
num2str(sqrt(mse(e)))])
```

```
98   display(['The RMSE of the testing is equal to '
99   num2str(sqrt(mse(et)))])
100
101  figure, plot(Modeling_Error,'k'); grid on; % The evolution of
102  % RMSE vs. epoch number
103  ylabel('The RMSE'); xlabel('Epoch');
104
105  T2FNN_output=inv_scale(T2FNN_output,norm_x(1),norm_x(2),
106  minn,maxn);
107  % Denormalization
108  [Data(1:row,col)]=inv_scale(Data(1:row,col),norm_x(1),
109  norm_x(2),minn,maxn);
110
111
112  figure, plot(Data(1:row,col),'k'); hold on;
113  plot(T2FNN_output,'k-.');
114  ylabel('The output')
115  xlabel('Sample')
116  legend('The measured output','The model output')
117
118  figure, plot(learning_rate,'k'); hold on;
119  ylabel('The learning rate alpha');
120  xlabel('Epoch');
121
122  figure
123  plot(qq,'k')
124  ylabel('The parameter (q)')
125  xlabel('Epoch')
```

A.1.2 Source Code for Matlab Function t2fnnsmc

```
1    function [yy,Cy,S_low_y,S_upp_y,ary,by,qy,alphay] =
2    t2fnnsmc(x,x_old,...
3        C,S_low,S_upp,ar,b,delta_t,q,alpha)
4    %%%%%%%%%%%%%%%%%%%%%%%%
5    %%% Initialization %%%
6    %%%%%%%%%%%%%%%%%%%%%%%%
7    [row1,col1]=size(C);
8    C_dot=zeros(row1,col1);
9    S_low_dot=zeros(row1,col1);
10   S_upp_dot=zeros(row1,col1);
11   alpha1=0.001;
12   %%%%%%%%%%%%%%%%%%%%%%%%%%%%%
13   %%% MF value calculation %%%
14   %%%%%%%%%%%%%%%%%%%%%%%%%%%%%%
15   for i=1:col1
16       M_low(:,i)=max(0,exp(-(x(i) - C(:,i)).^2./
```

```
    (2*S_low(:,i).^2)));
    M_upp(:,i)=max(0,exp(-(x(i) - C(:,i)).^2./
    (2*S_upp(:,i).^2)));
end
%%%%%%%%%%%%%%%%%%%%%%%%%%%%%%%%%%%%%%%%%%%%%%%%%
%%% The calcuation of the firing of the rules %%%
%%%%%%%%%%%%%%%%%%%%%%%%%%%%%%%%%%%%%%%%%%%%%%%%%
Wij_low = MatrixMultiple(M_low);
Wij_upp = MatrixMultiple(M_upp);
%%%%%%%%%%%%%%%%%%%%%%%%%%%%%%%%%%%%%%%%%%%%%%%%%%%%
%%% The normalization of the firing of the rules %%%
%%%%%%%%%%%%%%%%%%%%%%%%%%%%%%%%%%%%%%%%%%%%%%%%%%%%
W_low=Wij_low/(sum(Wij_low));
W_upp=Wij_upp/(sum(Wij_upp));
%%%%%%%%%%%%%%%%%%%%%%%%%%%%%%%%%%%%%%%%%%%%%%
%%% The calculation of the output of T2FNN %%%
%%%%%%%%%%%%%%%%%%%%%%%%%%%%%%%%%%%%%%%%%%%%%%
yy_low=(x(1:col1)*ar+b)*W_low;
yy_upp=(x(1:col1)*ar+b)*W_upp;
yy=q*yy_low+(1-q)*yy_upp;
e=yy-x(end); % The prediction error
signum=e/(abs(e)+0.05); % The approximate sign function
%%%%%%%%%%%%%%%%%%%%%%%%%%%%%%%%%%%%%%%%%%%%%
%%% Weight updates for constant parameters %%%
%%%%%%%%%%%%%%%%%%%%%%%%%%%%%%%%%%%%%%%%%%%%%
for i=1:row1^(col1)
   b_dot(1,i)=-alpha*signum*(q*W_low(i)+(1-q)*W_upp(i))...
   /(((q*W_low+(1-q)*W_upp)'*(q*W_low+(1-q)*W_upp)));
   by(1,i)=b(1,i)+b_dot(1,i);
end
%%%%%%%%%%%%%%%%%%%%%%%%%%%%%%%%%%%%%%%%%%%%%%%%%%%%%%%%%%%%%%

%%% Weight updates for the gains of the input in the %%%
%%% consequent part %%%
%%%%%%%%%%%%%%%%%%%%%%%%%%%%%%%%%%%%%%%%%%%%%%%%%%%%%%%%%%%%%%


for j=1:col1
   for i=1:row1^(col1)
      ar_dot(j,i)=-alpha*x(j)*signum*(q*W_low(i)+(1-q)*
      W_upp(i))...
      /(((q*W_low+(1-q)*W_upp)'*(q*W_low+(1-q)*W_upp)));
      ary(j,i)=ar(j,i)+ar_dot(j,i);
   end
end
%%%%%%%%%%%%%%%%%%%%%%%%%%%%%%%%%%%%%%%%%%
%%% Antecedent part parameters updates %%%
```

```
%%%%%%%%%%%%%%%%%%%%%%%%%%%%%%%%%%%%%%%%%%%
for j=1:col1
    for i=1:row1
        C_dot(i,j)=(x(i)-x_old(i))/delta_t+(x(i)-C(i,j))
        *alpha1*signum;
        % Adaptation laws of centers
        S_low_dot(i,j)=-(S_low(i,j)+S_low(i,j).^3/
        (x(i)-C(i,j)).^2) *alpha1*signum;
        % Adaptation laws of sigma lower
        S_upp_dot(i,j)=-(S_upp(i,j)+S_upp(i,j).^3/
        (x(i)-C(i,j)).^2) *alpha1*signum;
        % Adaptation laws of sigma upper
    end
end
Cy=C+C_dot*delta_t; % Update rule for centers
S_low_y=S_low+S_low_dot*delta_t; % Update rule for sigma
S_upp_y=S_upp+S_upp_dot*delta_t; % Update rule for sigma
q_dot =-delta_t*(alpha*signum)/(b*(W_low-W_upp));
qy=q+q_dot; % Update rule for q
% Restriction of q between '0' and '1'
if qy<0
    qy=0;
end
if qy>1
    qy=1;
end
% Update rule for the adaptive learning rate
alpha_dot =8e-6*abs(e)-1e-10*alpha;
alphay=alpha+alpha_dot;
```

A.1.3 Source Code for Matlab Function MatrixMultiple.m

```
function CC=MatrixMultiple(A)
% The calculation of the firing of the rules

m=size(A,2);

C=DoCalculation(A(:,m-1),A(:,m));
for i=1:(m-2)
    C=DoCalculation(A(:,m-(i+1)),C);
end
CC=C;

function C=DoCalculation(A,B)

k=size(B,1);
```

```
15  l=size(A,1);
16  C=zeros(l*k,1);
17  start=1; stop=start+k-1;
18  for i=1:l
19      C(start:stop,1)=A(i)*B;
20      start=start+k; stop=start+k-1;
21  end
```

A.1.4 Source Code for Matlab Function scale.m

```
1  function [xnew]=scale(xold,minold,maxold,minnew,maxnew)
2      xnew=minnew+(xold-minold)*(maxnew-minnew)/(maxold-minold);
```

A.1.5 Source Code for Matlab Function inv_scale.m

```
1  function [xnew]=inv_scale(xold,minold,maxold,minnew,maxnew)
2      xnew=(xold-minnew)*(maxold-minold)/(maxnew-minnew)+minold;
```

A.2 GD-BASED THEORY-BASED TRAINING ALGORITHM FOR TYPE-2 FUZZY NEURAL NETWORK

The main script file for the GD-based theory-based training algorithm for the consequent part parameters of T2FNN. In order to execute this script, several functions are needed. These functions are *"t2fnngd.m"*, "MatrixMultiple.m", "scale.m", and "inv_scale.m". The source code for these functions (except "t2fnngd.m") was given in the previous section, and the source code for "t2fnngd.m" is given after the main script file.

A.2.1 Main Script

```
1   clear all
2   close all
3   clc
4
5   %%%%%%%%%%%%%%%%%%%%%%%
6   %%% Data Generation %%%
7   %%%%%%%%%%%%%%%%%%%%%%%
8   x(1:18)=1.2; %Initial Conditions for Mackey-Glass
9   for t=18:2000
10      x(t+1)=x(t)-0.1*x(t)+0.2*x(t-17)/(x(t-17)^10+1); %The
11      % dynamic of Mackey-Glass
12  end
13  x=x(18:end);
```

```
%%%%%%%%%%%%%%%%%%%%%%%%%%
%%% Normalization Part %%%
%%%%%%%%%%%%%%%%%%%%%%%%%%
minn=0; maxn=1;
norm_x=minmax(x); y=scale(x,norm_x(1),norm_x(2),minn,maxn);
% Target normalization
%%%%%%%%%%%%%%%%%%%%%%%%%%%%%%%%%
%%% Data prepration for T2FNN %%%
%%%%%%%%%%%%%%%%%%%%%%%%%%%%%%%%%
tbeg=1; tend=length(y);
for t=2:1501
    Data(t,:)=[y(t) y(t+6) y(t+12) y(t+18) y(t+24)];
    % [y(t-18) y(t-12) y(t-6) y(t)] is the input and
    % y(t+6) is the target value of the T2FNN
end
[row,col]=size(Data);

%%%%%%%%%%%%%%%%%%%%%%%%%%%%
%%% T2FNN Initialization %%%
%%%%%%%%%%%%%%%%%%%%%%%%%%%%
q=0.5; % The initial condition for the q
center=[0.25 0.75]'; % The initial values of the centers of
% MFs for T2FNN
for i=1:col-1
    C(:,i)=center;
end
S_upp=0.3*ones(length(center),col-1);
% The initial values of the sigma upper of MFs for T2FNN
S_low=0.2*ones(length(center),col-1);
% The initial values of the sigma lower of MFs for T2FNN
ar=rand(col-1,length(center)^(col-1))*1e-2;
% The initial values of the gains of inputs in the
% consequent part
b=rand(1,length(center)^(col-1))*1e-2;
% The initial values of the constant terms in the
% consequent part

alpha=0.1; % The learning rate

%%%%%%%%%%%%%%%%%
%%% Main loop %%%
%%% Training %%%
%%%%%%%%%%%%%%%%%
for epoch=1:100
    for i=3:1000
```

```
        [T2FNN_output(i),ar,b,q] = t2fnngd(Data(i,:),C,S_low,
        S_upp,ar,b,q,alpha);
        % The calculation of the output of T2FNN and its
        % update procedure
        % based on GD training algorithm

        e(i-2)=Data(i,col)-T2FNN_output(i); %The training
            error
    end
    qq(epoch)=q; %The updated value of q
    Modeling_Error(epoch)=sqrt(mse(e));
    %The value of the RMSE obtained at the end of each epoch
end


%%%%%%%%%%%%%%%
%%% Testing %%%
%%%%%%%%%%%%%%%
for i=1001:1200
      T2FNN_output(i) = t2fnngd(Data(i,:),C,S_low,S_upp,ar,
      b,q,alpha);
      %Output evaluation

      et(i-1000)=T2FNN_output(i)-Data(i,col);
end

%%%%%%%%%%%%%%%
%%% Display %%%
%%%%%%%%%%%%%%%
display(['The RMSE of the training is equal to '
num2str(sqrt(mse(e)))])
display(['The RMSE of the testing is equal to '
num2str(sqrt(mse(et)))])

figure, plot(Modeling_Error,'k'); grid on; % The evolution
% of RMSE vs. epoch number
ylabel('The RMSE'); xlabel('Epoch');

T2FNN_output=inv_scale(T2FNN_output,norm_x(1),norm_x(2),
minn,maxn);
% Denormalization
[Data(1:row,col)]=inv_scale(Data(1:row,col),norm_x(1),
norm_x(2),minn,maxn);


figure, plot(Data(1:row,col),'k'); hold on;
plot(T2FNN_output,'k-.');
```

```
107  ylabel('The output')
108  xlabel('Sample')
109  legend('The measured output','The model output')
110
111
112  figure
113  plot(qq,'k')
114  ylabel('The parameter (q)')
115  xlabel('Epoch')
```

A.2.2 Source Code for Matlab Function t2fnngd.m

```
1   function [yy,ary,by,qy] = t2fnngd(x,C,S_low,S_upp,ar,b,q,
        alpha)
2   %%%%%%%%%%%%%%%%%%%%%%%
3   %%% initialization %%%
4   %%%%%%%%%%%%%%%%%%%%%%%
5   [row1,col1]=size(C);
6   %%%%%%%%%%%%%%%%%%%%%%%%%%%%%
7   %%% MF value calculation %%%
8   %%%%%%%%%%%%%%%%%%%%%%%%%%%%%
9   for i=1:col1
10      M_low(:,i)=max(0,exp(-(x(i) - C(:,i)).^2./
11      (2*S_low(:,i).^2)));
12      M_upp(:,i)=max(0,exp(-(x(i) - C(:,i)).^2./
13      (2*S_upp(:,i).^2)));
14  end
15  %%%%%%%%%%%%%%%%%%%%%%%%%%%%%%%%%%%%%%%%%%%%%%%%%
16  %%% The calcuation of the firing of the rules %%%
17  %%%%%%%%%%%%%%%%%%%%%%%%%%%%%%%%%%%%%%%%%%%%%%%%%
18  Wij_low = MatrixMultiple(M_low);
19  Wij_upp = MatrixMultiple(M_upp);
20  %%%%%%%%%%%%%%%%%%%%%%%%%%%%%%%%%%%%%%%%%%%%%%%%%%
21  %%% The normalization of the firing of the rules %%%
22  %%%%%%%%%%%%%%%%%%%%%%%%%%%%%%%%%%%%%%%%%%%%%%%%%%
23  W_low=Wij_low/(sum(Wij_low));
24  W_upp=Wij_upp/(sum(Wij_upp));
25  %%%%%%%%%%%%%%%%%%%%%%%%%%%%%%%%%%%%%%%%%%
26  %%% The calculation of the output of T2FNN %%%
27  %%%%%%%%%%%%%%%%%%%%%%%%%%%%%%%%%%%%%%%%%%
28  yy_low=(x(1:col1)*ar+b)*W_low;
29  yy_upp=(x(1:col1)*ar+b)*W_upp;
30  yy=q*yy_low+(1-q)*yy_upp;
31  e=x(end)-yy; % The prediction error
32  %%%%%%%%%%%%%%%%%%%%%%%%%%%%%%%%%%%%%%%%%%
33  %%% Weight updates for constant parameters %%
```

```
%%%%%%%%%%%%%%%%%%%%%%%%%%%%%%%%%%%%
for i=1:row1^(col1)
    delta_b(1,i)=alpha*e*(q*W_low(i)+(1-q)*W_upp(i));
    by(1,i)=b(1,i)+delta_b(1,i);
end
%%%%%%%%%%%%%%%%%%%%%%%%%%%%%%%%%%%%%%%%%%%%%%%%%%%%%%%%%%%%
%%% Weight updates for the gains of the input in the
%%% consequent part %%%
%%%%%%%%%%%%%%%%%%%%%%%%%%%%%%%%%%%%%%%%%%%%%%%%%%%%%%%%%%%%
for j=1:col1
    for i=1:row1^(col1)
        delta_ar(j,i)=alpha*e*x(j)*(q*W_low(i)+(1-q)*W_upp(i))
            ;
        ary(j,i)=ar(j,i)+delta_ar(j,i);
    end
end

qy=q+alpha*e*(yy_low-yy_upp);
if qy>1
    qy=1;
elseif qy<0
    qy=0;
end
```

A.3 KF-BASED TRAINING ALGORITHM FOR TYPE-2 FUZZY NEURAL NETWORK

The main script file for KF-Based theory-based training algorithm for the parameters of the consequent part of T2FNN. In order to execute this script, several functions are needed. These functions are "t2fnnkalman.m", "MatrixMultiple.m", "scale.m", and "inv_scale.m". The source code for these functions except "t2fnngd.m" was given in the previous sections, and the source code for "t2fnnkalman.m" is given after the main script file.

A.3.1 Main Script

```
clear all
close all
clc

%%%%%%%%%%%%%%%%%%%%%%
%%% Data Generation %%%
%%%%%%%%%%%%%%%%%%%%%%
x(1:18)=1.2; %Initial Conditions for Mackey-Glass
for t=18:2000
```

```
    x(t+1)=x(t)-0.1*x(t)+0.2*x(t-17)/(x(t-17)^10+1); %The
    % dynamic of Mackey-Glass
end
x=x(18:end);

%%%%%%%%%%%%%%%%%%%%%%%%%%
%%% Normalization Part %%%
%%%%%%%%%%%%%%%%%%%%%%%%%%
minn=0; maxn=1;
norm_x=minmax(x); y=scale(x,norm_x(1),norm_x(2),minn,maxn);
% Target normalization
%%%%%%%%%%%%%%%%%%%%%%%%%%%%%%%%
%%% Data prepration for T2FNN %%%
%%%%%%%%%%%%%%%%%%%%%%%%%%%%%%%%
tbeg=1; tend=length(y);
for t=2:1501
    Data(t,:)=[y(t) y(t+6) y(t+12) y(t+18) y(t+24)];
    % [y(t-18) y(t-12) y(t-6) y(t)] is the input and
    % y(t+6) is the target value of the T2FNN
end
[row,col]=size(Data);

%%%%%%%%%%%%%%%%%%%%%%%%%%%%
%%% T2FNN Initialization %%%
%%%%%%%%%%%%%%%%%%%%%%%%%%%%
q=0.5; % The initial condition for q
 center=[0.25 0.75]'; % The initial values of the centers of
 % MFs for T2FNN
for i=1:col-1
   C(:,i)=center;
end
S_upp=0.3*ones(length(center),col-1);
% The initial values of the sigma upper of MFs for T2FNN
S_low=0.2*ones(length(center),col-1);
% The initial values of the sigma lower of MFs for T2FNN
ar=rand(col-1,length(center)^(col-1))*1e-2;
% The initial values of the gains of inputs in the consequent
% part
b=rand(1,length(center)^(col-1))*1e-2;
% The initial values of the constant terms in the consequent
% part

P=eye(size(ar,1)*size(ar,2)+length(b)); % The covariance
     matrix
R=1; % The covariance of measurement noise
Q=0.001*eye(size(ar,1)*size(ar,2)+length(b)); % The
     covariance
```

```
% of process noise
%%%%%%%%%%%%%%%%%
%%% Main loop %%%
%%% Training %%%
%%%%%%%%%%%%%%%%%
for epoch=1:100
    for i=3:1000
        [T2FNN_output(i),ar,b,P] = t2fnnkalman(Data(i,:),
        C,S_low,S_upp,...
        ar,b,q,P,Q,R);
        % The calculation of the output of T2FNN and its
        % update procedure
        % based on KF training algorithm
        e(i-2)=Data(i,col)-T2FNN_output(i); %The training error
    end
    Modeling_Error(epoch)=sqrt(mse(e)); %The value of the RMSE
    % obtained at the end of each epoch
end
%%%%%%%%%%%%%%%%
%%% Testing %%%
%%%%%%%%%%%%%%%%
for i=1001:1200
    T2FNN_output(i) = t2fnnkalman(Data(i,:),C,S_low,S_upp,
    ar,b,q,P,Q,R);
    %Output evaluation
    et(i-1000)=T2FNN_output(i)-Data(i,col);
end
%%%%%%%%%%%%%%%%
%%% Display %%%
%%%%%%%%%%%%%%%%
display(['The RMSE of the training is equal to '
num2str(sqrt(mse(e)))])
display(['The RMSE of the testing is equal to '
num2str(sqrt(mse(et)))])
figure, plot(Modeling_Error,'k'); grid on; % The evolution of
RMSE vs. epoch number
ylabel('The RMSE'); xlabel('Epoch');
```

```
T2FNN_output=inv_scale(T2FNN_output,norm_x(1),
norm_x(2),minn,maxn);
% Denormalization
[Data(1:row,col)]=inv_scale(Data(1:row,col),norm_x(1),
norm_x(2),minn,maxn);


figure, plot(Data(1:row,col),'k'); hold on;
plot(T2FNN_output,'k-.');
ylabel('The output')
xlabel('Sample')
legend('The measured output','The model output')
```

A.3.2 Source Code for Matlab Function t2fnnkalman.m

```
function [yy,ary,by,P] = t2fnnkalman(x,C,S_low,S_upp,ar,
b,q,P,Q,R)
%%%%%%%%%%%%%%%%%%%%%%
%%% initialization %%%
%%%%%%%%%%%%%%%%%%%%%%
[row1,col1]=size(C);
%%%%%%%%%%%%%%%%%%%%%%%%%%%%
%%% MF value calculation %%%
%%%%%%%%%%%%%%%%%%%%%%%%%%%%
for i=1:col1
    M_low(:,i)=max(0,exp(-(x(i) - C(:,i)).^2./
    (2*S_low(:,i).^2)));
    M_upp(:,i)=max(0,exp(-(x(i) - C(:,i)).^2./
    (2*S_upp(:,i).^2)));
end
%%%%%%%%%%%%%%%%%%%%%%%%%%%%%%%%%%%%%%%%%%%%%%%%%
%%% The calcuation of the firing of the rules %%%
%%%%%%%%%%%%%%%%%%%%%%%%%%%%%%%%%%%%%%%%%%%%%%%%%
Wij_low = MatrixMultiple(M_low);
Wij_upp = MatrixMultiple(M_upp);
%%%%%%%%%%%%%%%%%%%%%%%%%%%%%%%%%%%%%%%%%%%%%%%%%%%%
%%% The normalization of the firing of the rules %%%
%%%%%%%%%%%%%%%%%%%%%%%%%%%%%%%%%%%%%%%%%%%%%%%%%%%%
W_low=Wij_low/(sum(Wij_low));
W_upp=Wij_upp/(sum(Wij_upp));
%%%%%%%%%%%%%%%%%%%%%%%%%%%%%%%%%%%%%%%%%%%%%%
%%% The calculation of the output of T2FNN %%%
%%%%%%%%%%%%%%%%%%%%%%%%%%%%%%%%%%%%%%%%%%%%%%
yy_low=(x(1:col1)*ar+b)*W_low;
yy_upp=(x(1:col1)*ar+b)*W_upp;
yy=q*yy_low+(1-q)*yy_upp;
e=x(end)-yy; % The prediction error
```

```
%%%%%%%%%%%%%%%%%%%%%%%%%%%%%%%%%%%%%%%%%%%%%
%%% Weight updates for constant parameters %%
%%%%%%%%%%%%%%%%%%%%%%%%%%%%%%%%%%%%%%%%%%%%%
Phi=q*W_low+(1-q)*W_upp;
%%%%%%%%%%%%%%%%%%%%%%%%%%%%%%%%%%%%%%%%%%%%%%%%%%%%%%%%%%%%%
%%% Weight updates for the gains of the input in the
%%% consequent part %%%
%%%%%%%%%%%%%%%%%%%%%%%%%%%%%%%%%%%%%%%%%%%%%%%%%%%%%%%%%%%%%
for j=1:coll
    Phi=[Phi;(q*W_low+(1-q)*W_upp)*x(j)];
end

L=P*Phi*(Phi'*P*Phi+R)^(-1);

P=P-L*Phi'*P+Q; % The update rule for covariance matrix
d=L*e;

lar=size(ar,2);
by=(b+d(1:lar,1)');% Adaptation for b
for j=2:size(ar,1)+1
    ary(j-1,:)=(ar(j-1,:)+d((j-1)*lar+1:j*lar,1)');%
        Adaptation
    % for ar
end
```

Appendix B

Contents

Abstract

This chapter provides sample source code for the T2FNN training algorithm used as controller. The programming language is MATLAB.

Keywords

Simulation, Controller, MATLAB source code

The main script file for a full SMC theory-based algorithm for the training of a T2FNN when used as a controller is given here. In order to execute this script, two functions are needed: *neurofuzzy.m* and *elip.m*. The source code for these functions must be written as an independent file. This code is provided after the main script.

B.1 SMC THEORY-BASED TRAINING ALGORITHM FOR TYPE-2 FUZZY NEURAL NETWORK CONTROLLER

B.1.1 Main Script

```
close all;clear all;clc
%% The plant parameters%%
alpha1=0.6984;K=1.1;gamma=0.2;
%% The initial conditions of the system
x1=0.6;x2=0.1;
%% T2FNN configuration
c1=[-1 0 1];
c2=[-1 0 1];
d1=ones(1,3);
d2=ones(1,3);
a11=d1;a21=d1;a12=d1;a22=d1;q=0.5;
theta=zeros(3,3);kk=1;
%% Simulation parameters
T=20;
```

```
Ts=0.001;
%% The parameters of the PD controller
Kp=9;Kd=5;u=0;
%% The learning rates
gamma1=0.01;alpha=4;
%% The reference signal
rr=0.4*sin(kk*(1:T/Ts+10)*Ts);Kr=0;alpha=0;
for t=1:T/Ts
   %% The dynamic model of the system
   x1(t+1)=x1(t)+Ts*x2(t);
   x2(t+1)=x2(t)+Ts*(-K*sin(2*x1(t))-gamma*x2(t)-alpha1*(2*sin(x1(t))*...
      sin(t*Ts)+cos(x1(t))*cos(t*Ts))+u(t));
   %% The classical controller calculations
   e=rr(t+1)-x1(t+1);
   de=0.4*kk*cos(kk*t*Ts)-x2(t+1);
   tauc(t+1)=Kp*(e)+Kd*(de);
   %% The T2FNN controller calculations
   [y theta c1 c2 d1 d2 a11 a21 a12 a22]=neurofuzzy1(e,de,q,tauc(t+1),...
      theta,c1,c2,d1,d2,a11,a21,a12,a22,gamma1,alpha(t),Ts);
   %%The overall control signal
   u(t+1)=tauc(t+1)-y+Kr(t)*de/(abs(de)+0.01);
   %% The adaptive parameter of the term to guarantee the robustness
   Kr(t+1)=Kr(t)+0.001*abs(de);
   %% The adaptive learning rate
   alpha(t+1)=alpha(t)+0.001*abs(tauc(t+1));
   %% The error signal
   ee(t)=e;
end

figure(1)
plot((0:T/Ts)*Ts,x1)
hold on
figure(4)
plot((0:T/Ts)*Ts,tauc)
hold on
plot((0:T/Ts)*Ts,-u+tauc,'r')
xlabel('t')
plot((0:T/Ts)*Ts,u,'k')
legend('The output of the PD controller',...
   'The output of the type-2 fuzzy controller','The overall control signal')
ylabel('The control signals')
%% The simulations corresponding to the PD controller when acting alone
```

```
for t=1:T/Ts

    x1(t+1)=x1(t)+Ts*x2(t);
    x2(t+1)=x2(t)+Ts*(-K*sin(2*x1(t))-gamma*x2(t)-alpha1*(2*sin(x1(t))*...
      sin(t*Ts)+cos(x1(t))*cos(t*Ts))+u(t));

    e=rr(t+1)-x1(t+1);
    de=0.4*kk*cos(kk*t*Ts)-x2(t+1);
    tauc(t+1)=Kp*(e)+Kd*(de);
    u(t+1)=tauc(t+1);
    ee(t)=e;
end

figure(1)
plot((0:T/Ts)*Ts,x1,'r')
TT1=(0:T/Ts)*Ts;
RR1=rr(1:20001);
plot(TT1(1:500:20001),RR1(1:500:20001),'k')
ylabel('x_1')
axis([0 20 -0.6 1.6])
xlabel('t')
legend('The response of the proposed control system',...
    'The response of the PD controller when it is used alone',...
    'The sinusoidal reference signal')
figure(2)
plot((0:T/Ts)*Ts,Kr,'k')
xlabel('t')
ylabel('The adaptation of the parameter to guarantee robustness (K_r)')
figure(3)
plot((0:T/Ts)*Ts,alpha,'k')
xlabel('t')
ylabel('The adaptation of the learning rate (\alpha)')
```

B.1.2 Source Code for MATLAB Function neurofuzzy

```
function [y thetan c1 c2 d1 d2 a11 a21 a12 a22]=neurofuzzy(e,de,q,tauc,...
    theta,c1,c2,d1,d2,a11,a21,a12,a22,gamma1,alpha,Ts)

%% The calculation of the upper and lower MF for e
for i=1:length(c1)
    mho1_lower(i)=elip(e,c1(i),d1(i),a21(i));
    mho1_upper(i)=elip(e,c1(i),d1(i),a11(i));
end
```

```
%% The calculation of the upper and lower MF for de
for i=1:length(c2)
   mho2_lower(i)=elip(de,c2(i),d2(i),a22(i));
   mho2_upper(i)=elip(de,c2(i),d2(i),a12(i));
end
%% The calculation of the upper and lower and upper bounds of the rules
for i=1:length(c1)
   for j=1:length(c2)
      rule_lower(i,j)=mho1_lower(i)*mho2_lower(j);
      rule_upper(i,j)=mho1_lower(i)*mho2_lower(j);
   end
end
%% The normalization of the rules
rule_lower_tilde=rule_lower*sum(sum(rule_lower))/(sum(sum(rule_lower)...
   )^2
+0.001);
rule_upper_tilde=rule_upper*sum(sum(rule_upper))/(sum(sum(rule_upper)...
   )^2+0.001);
y=sum(sum(rule_lower_tilde.*theta*q+rule_upper_tilde.*theta*(1-q)));


if sum(sum(rule_upper))+ sum(sum(rule_lower)) ==0
   y=0;
   thetan=theta;
else

  %% The adaptation rules
   for i=1:length(a21)
      T21(i)=(abs((e-c1(i))/d1(i)))^a21(i);
      T11(i)=(abs((e-c1(i))/d1(i)))^a11(i);
      T22(i)=(abs((de-c2(i))/d2(i)))^a22(i);
      T12(i)=(abs((de-c2(i))/d2(i)))^a12(i);
      if e<=c1(i)-d1(i)
         a21(i)=a21(i);
      elseif e>=c1(i)+d1(i)
         a21(i)=a21(i);
      elseif T21(i)==0
         a21(i)=a21(i);
      else
         a21(i)=a21(i)+Ts*gamma1*(log(1-(abs((e-c1(i))/d1(i)))^a21(i)...
            )/a21(i)^2+(abs((e-c1(i))...
```

```
            /d1(i)))^a21(i)*log(abs((e-c1(i))/d1(i)))/a21(i))^(-1)*...
            (1-T21(i))*H(e,c1(i)-d1(i),c1(i)+d1(i));
    end
    if e<=c1(i)-d1(i)
        a11(i)=a11(i);
    elseif e>=c1(i)+d1(i)
        a11(i)=a11(i);
    elseif T11(i)==0
        a11(i)=a11(i);
    else
        a11(i)=a11(i)+Ts*gamma1*(log(1-(abs((e-c1(i))/d1(i)))^a11(i)...
            )/a11(i)^2+(abs((e-c1(i))...
            /d1(i)))^a11(i)*log(abs((e-c1(i))/d1(i)))/a11(i))^(-1)...
            *(1-T11(i))*H(e,c1(i)-d1(i),c1(i)+d1(i));
    end
    if de<=c2(i)-d2(i)
        a22(i)=a22(i);
    elseif de>=c2(i)+d2(i)
        a22(i)=a22(i);
    elseif T22(i)==0
        a22(i)=a22(i);
    else
        a22(i)=a22(i)+Ts*gamma1*(log(1-(abs((de-c2(i))/d2(i)))^a22...
            (i))/a22(i)^2+(abs((de-c2(i))...
            /d2(i)))^a22(i)*log(abs((de-c2(i))/d2(i)))/a22(i))^(-1)...
            *(1-T22(i))*H(de,c2(i)-d2(i),c2(i)+d2(i));
    end
    if de<=c2(i)-d2(i)
        a12(i)=a12(i);
    elseif e>=c2(i)+d2(i)
        a12(i)=a12(i);
    elseif T12(i)==0
        a12(i)=a12(i);
    else
        a12(i)=a12(i)+Ts*gamma1*(log(1-(abs((de-c2(i))/d2(i)))^a12...
            (i))/a12(i)^2+(abs((de-c2(i))...
            /d2(i)))^a12(i)*log(abs((de-c2(i))/d2(i)))/a12(i))^(-1)...
            *(1-T12(i))*H(de,c2(i)-d2(i),c2(i)+d2(i));
    end
    if a11(i)<1
```

```
            a11(i)=1;
        end
        if a21(i)>1
            a21(i)=1;
        end
        if a12(i)<1
            a21(i)=1;
        end
        if a22(i)>1
            a22(i)=1;
        end
    end

    for i=1:length(c1)
        if e-c1(i) ~=0
            c1(i)=c1(i)-gamma1*Ts*abs(d1(i))^a21(i)*(1-T21(i))*
                sign(e-...
                c1(i))/(abs(e-c1(i)))^(a21(i)-1)*H(e,c1(i)-d1(i),
                    c1(i)...
                +d1(i)) ;
        end
        if de-c2(i) ~=0
            c2(i)=c2(i)-gamma1*Ts*abs(d2(i))^a22(i)*(1-T22(i))*
                sign(de-...
                c2(i))/(abs(de-c1(i)))^(a22(i)-1)*H(de,c2(i)-d2(i
                    ),c2(i)...
                +d2(i));
        end
        if e-c1(i) ~=0
            d1(i)=d1(i)-gamma1*Ts*(1-T21(i))*abs(d1(i))^(a21(i)
                +1)/(abs...
                (e-c1(i)))^(a21(i))*sign(d1(i))*H(e,c1(i)-d1(i),
                    c1(i)+d1(i));
        end
        if de-c2(i) ~=0
            d2(i)=d2(i)-gamma1*Ts*(1-T22(i))*abs(d2(i))^(a22(i)
                +1)/(abs(...
                de-c2(i)))^(a22(i))*sign(d2(i))*H(de,c2(i)-d2(i),
                    c2(i)...
                +d2(i));
        end
    end


    for i=1:length(c1)
        for j=1:length(c2)
```

```
            theta(i,j)=theta(i,j)-Ts*(q*rule_lower_tilde(i,j)+(1-q)*...
                rule_upper_tilde(i,j))...
                /sum(sum(q*rule_lower_tilde+(1-q)*rule_upper_tilde))/...
                sum(sum(q*rule_lower_tilde+(1-q)*rule_upper_tilde))*...
                alpha*tauc/(abs(tauc)+0.001);
        end
    end
    thetan=theta;
end

function y=H(x,m,n)
if x<m
    y=0;
elseif x>n
    y=0;
else
    y=x;
end
```

B.1.3 Source Code for MATLAB Function elip.m

```
function y=elip(x,c,d,a)

if x<=c-d
    y=0;
elseif x>=c+d
    y=0;
else
    y=(1-(abs((x-c)/d))^a)^(1/a);
end
```

INDEX

Note: Page numbers followed by *f* indicate figures and *t* indicate tables.

Printed in the United States
By Bookmasters